The Quantum World

and the
Beginning of the Universe

David Chapple

Published 2012 by abramis

ISBN 978 1 84549 555 8

Printed and bound in the United Kingdom

Typeset in Garamond

abramis is an imprint of arima publishing.

arima publishing
ASK House, Northgate Avenue
Bury St Edmunds, Suffolk IP32 6BB
t: (+44) 01284 700321

www.abramis.co.uk

Foreword

What is quantum physics?

Quantum physics is the physics of sub atomic particles, and, unlike the predictable format of classical physics, quantum physics appears to be totally indeterminate, which makes it a very exciting subject to study.

In this book, we look at the energy associated with quantum particles, usually photons of electromagnetic radiation.

We then examine how to determine the distance between electron shells in hydrogen atoms, and finally we consider the Rydberg constant which will enable us to calculate the wavelength of the photons that can be emitted from an electron collapse between shells.

We also consider the time independent form of the Schrödinger equation in order to examine the phenomenon of quantum tunnelling.

We then consider the philosophical side of quantum theory, not least of which is the multiple state theory, which, of course, is mathematically faultless.

We complete our book by examining the possibility that the whole Universe developed all possible histories and when, so called, intelligent life evolved, the finely tuned constants of the Universe were locked in our part of the history.

Chapter 1:   Introduction

What is the quantum physics?

It is generally regarded as having two distinct thrusts.

One thrust is called in some circles **"quantum cookery"**.

What does this mean?

It simply means that some quantum physicists are only concerned with the equations of the quantum physics, and it must be said that the success of the quantum physics equations are profoundly successful.

The other thrust is **"quantum theory, or quantum philosophy"**.

What is this?

This involves the effort by some to try to understand how the quantum world works.

It must be said that the ideas as to how the quantum world works can be interpreted by some as extremely counterintuitive.

In the early great days of the discoveries of the quantum equations (usually taken to be in the 1920s), the general philosophy was to use the equations and forget about how the quantum world worked.

However, today there are many quantum physicists who are very interested in examining and experimenting with the quantum world to get some sort of grasp as to how it works, although it must be said that to date at the time of writing, we do not have a clear grasp of how nor why the strange events that we shall talk about in this book come to be!

For example, we will be asking if quantum particles can possibly exhibit some form of consciousness, and we will be examining the possibility that quantum particles might be able to traverse time, as a wave.

We will also be mystified at the ability of quantum particles to 'know' when they are being observed, and then to see them refuse to act as waves, but always to act as solid particles.

The question we will want to answer (although there are no answers at this time), is how can quantum particles possibly know that somebody is trying to observe them – because they can be uncannily aware of any attempt to observe them, and how they know is a complete mystery.

In fact, the double slit experiment and its many strange observations is quite impossible to explain in any sort of classical physics way.

The behaviour of quantum particles when they are being observed is completely impossible to explain with our present understanding of the quantum physical world, as we will soon see.

Chapter 2:   The Planck Constant

In the approach to the year 1900, many physicists were thinking about blackbody radiation.

What is this?

The problem was related to the continuous spectrum of light emitted from hot objects.

A typical spectrum is as shown below:

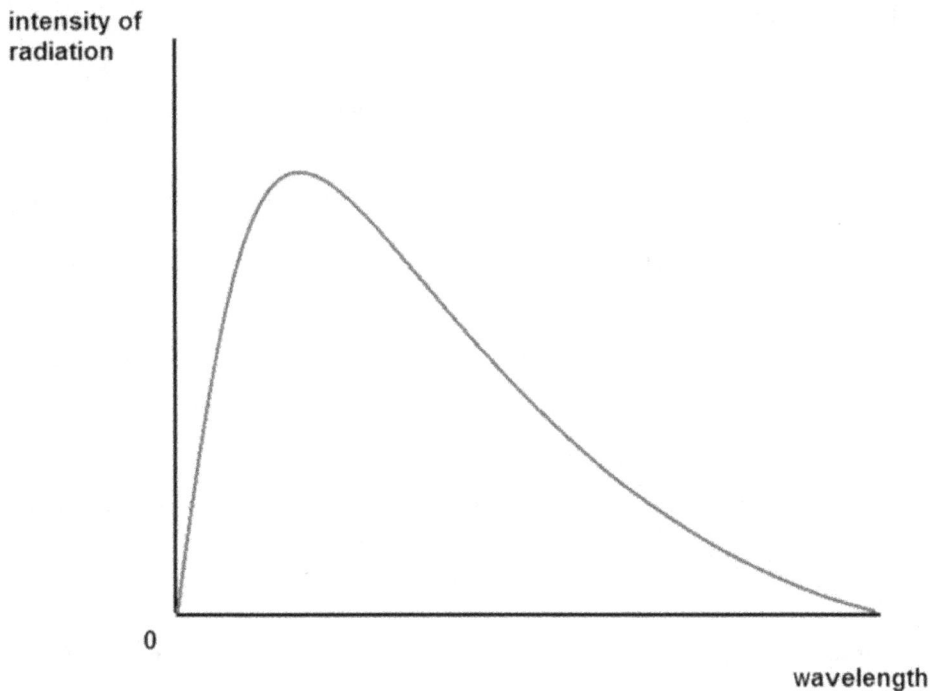

intensity of
radiation

0

wavelength

When an object is heated to some hundreds of degrees, it begins to glow.

The radiation emitted covers a range of wavelengths, part of which might lie within the visible range.

Some observed colours together with their approximate temperatures are shown below:

Colour of heated object	Temperature (degrees celsius)
red	500 - 950
yellow	1050 - 1200
white	1200 - 1600

The change in the objects observed colour can be used to estimate its temperature.

The ideal object name for absorption and radiation of radiation is called a *blackbody*.

A blackbody is such that at low temperatures, the body would absorb *all* the radiation which fell upon it.

And at higher temperatures it would emit all its radiation, so in effect it is a theoretical perfect absorber and radiator.

It is well known that for any temperature T, one wavelength has a greater intensity than all the others.

Now, the *Wien Displacement Law* can give us the wavelength compared with the temperature.

This law is given here:

$$\lambda_m T \ = \ 2.90 \ x \ 10^{-3} mK$$

As a result, the spectra for a body at different temperatures can be as described below:

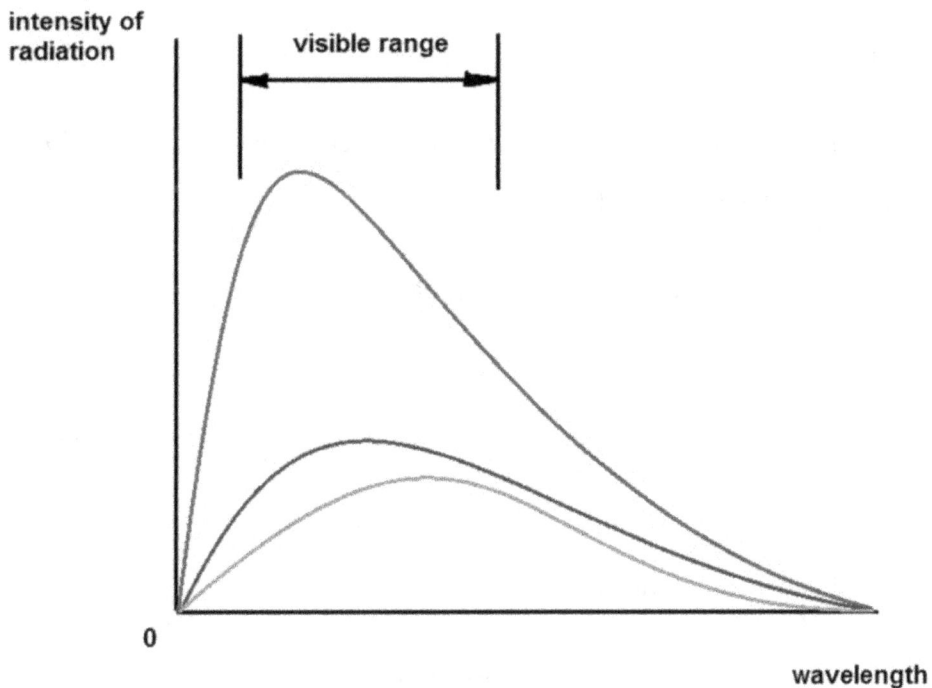

Notice that the colours in the diagram do not represent the true colours of the radiation.

Notice also that as the temperature increases, the wavelength of the radiation decreases.

Many attempts were made from classical physics to explain the shape of such spectra.

However, they all failed totally to reproduce the experimentally observed results.

Then, in 1900, Max Planck published his discovery of an equation which did, in fact, describe the experimental results perfectly.

However, this was not a classical physics equation.

Planck had, for the first time, realised that the spectra were explained by *quantum theory* as opposed to classical physics.

Planck had for the first time postulated that the energy in the spectra was *quantised*, such that it was limited to certain discrete values.

Planck had discovered that the energy of spectra was, in fact, proportional to it *frequency*.

The Planck equation describing this is given as:

$$E = hf$$

The wave equation enables us to reduce this to:

$$E = \frac{h\,c}{\lambda}$$

In the equation, **h** is referred to as Planck's constant, and it has the value of

6.626070×10^{-34}Js

The units are identical to the units of angular momentum.

However, if we wish to express the frequency in terms of radians per second, as opposed to cycles per second, we then use what is called the *reduced Planck constant*, sometimes referred to as the *Dirac constant*.

This is quoted as $\hbar$ instead of just **h**.

The reduced Planck constant is given as:

$$\hbar = \frac{h}{2\pi}$$

The value of the reduced Planck constant is given as **1.054571×10^{-34}Js**

So just what is the Planck constant?

It is generally regarded as the constant of the very small – in other words, it is regarded by a lot of physicists as the *quantum constant*, since it appears regularly in association with the quantum physics.

It was later that Einstein continued work with what Planck had started, and he achieved his photoelectric equation, which may be used to calculate whether or not a photon can ionise an electron and it so, how much energy would be transferred to the ionised photoelectron.

There are many instances where the Planck constant is used to perform calculations, especially in the quantum realm.

Even length and time are grainy in our Universe.

The Planck length is actually the smallest length that is possible in this Universe.

Quantum chronodynamics tells us that the Planck length is calculated from the following expression, in which the reduced Planck constant is deployed:

$$\lambda_p = \sqrt{\frac{G\,\hbar}{c^3}}$$

In the above equation, $\hbar$ is referred to as the **reduced Planck constant** and is calculated from

$$\frac{h}{2\pi}$$

Its value is given as **1.054593×10^{-34}Js**

Calculate now the smallest possible length permissible in this Universe.

Check your answer in the answer chapter, chapter 20, at the end of the book **(h)**.

It is impossible to speak of any length less than this figure.

Similarly, the Planck time is the smallest possible time it is possible to have in this Universe – even time is grainy.

This means that time does not flow smoothly, but even time itself is quantised.

The Planck time is defined as the time taken for a photon to pass through the Planck distance.

Calculate this smallest possible time now, and check your answer in chapter 20 at the end of the book.

It is impossible to conceive of any time interval smaller than this.

Chapter 3: The Einstein Contribution

Many people think that the great physicist Albert Einstein was awarded his Nobel Prize for his work on Special and General Relativity, because it is this that he is most well known for.

However, he actually was awarded his Nobel Prize in 1922 for his work on quantum physics, which he then spent the remaining years of his life opposing.

He, of course, did not oppose his own findings, but he was very unhappy about the explanations as to how the quantum world works, which, all are agreed, is very counterintuitive.

He, Boris Podolsky and Nathan Rosen devised the famous EPR thought experiment which was designed to throw a challenge to the thinking of the time as to how the quantum world worked, in particular, quantum entanglement, which Einstein called "spooky action at a distance".

We will examine this "gedankenexperiment" in some detail later.

The Nobel Prize in Physics was awarded to Albert Einstein "for his services to Theoretical Physics and especially for his discovery of the law of the photoelectric effect"

So what exactly is the photoelectric effect?

Let's find out.

The prevailing model of the atom at the time was the Rutherford model.

This claimed that the electron was a solid particle (which is also a worthy view of the electron as well as the wave model), and the electron 'orbited' the nucleus.

It was something like this:

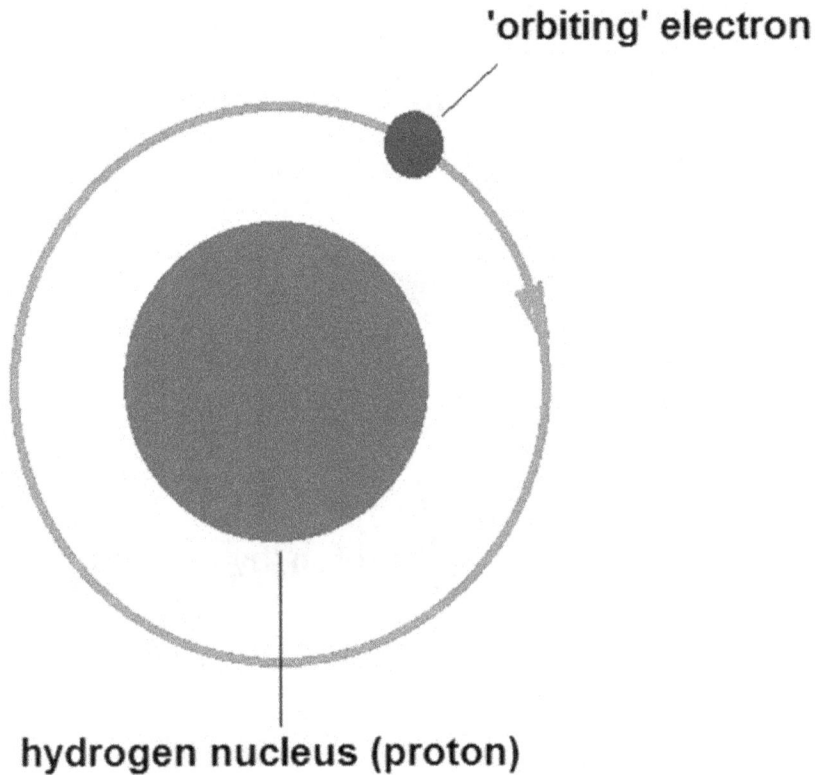

The problem was that an orbiting electron is continually accelerating (remember that a change in **direction** always results in an acceleration as well as a change in **speed**, since velocity is a vector quantity.

Now, any accelerating charged particle will emit energy in the form of electromagnetic radiation.

This should result in the electron 'spiralling' in towards the nucleus, something like this for the hydrogen atom:

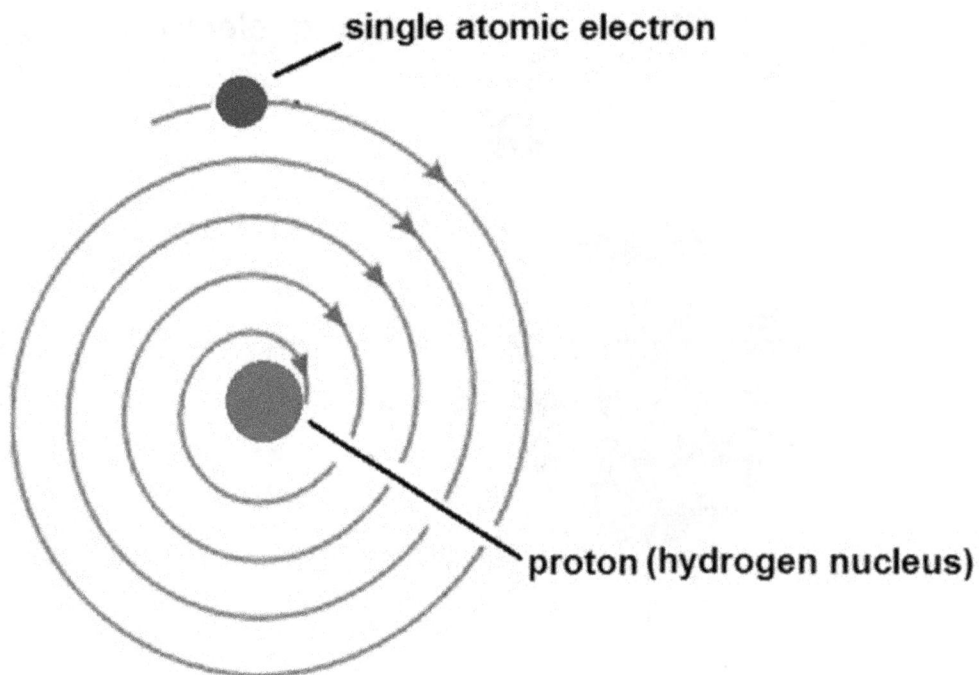

single atomic electron

proton (hydrogen nucleus)

Clearly this does not happen, or else we would be 'atomless', and clearly we are not.

This means that the atoms in our bodies are stable and the electrons do not spiral in towards the nuclei.

So a new model of the atom was needed.

This model came from the famous quantum physicist Niels Bohr, and is known as the 'Bohr model of the atom'.

For hydrogen, it can be represented like this:

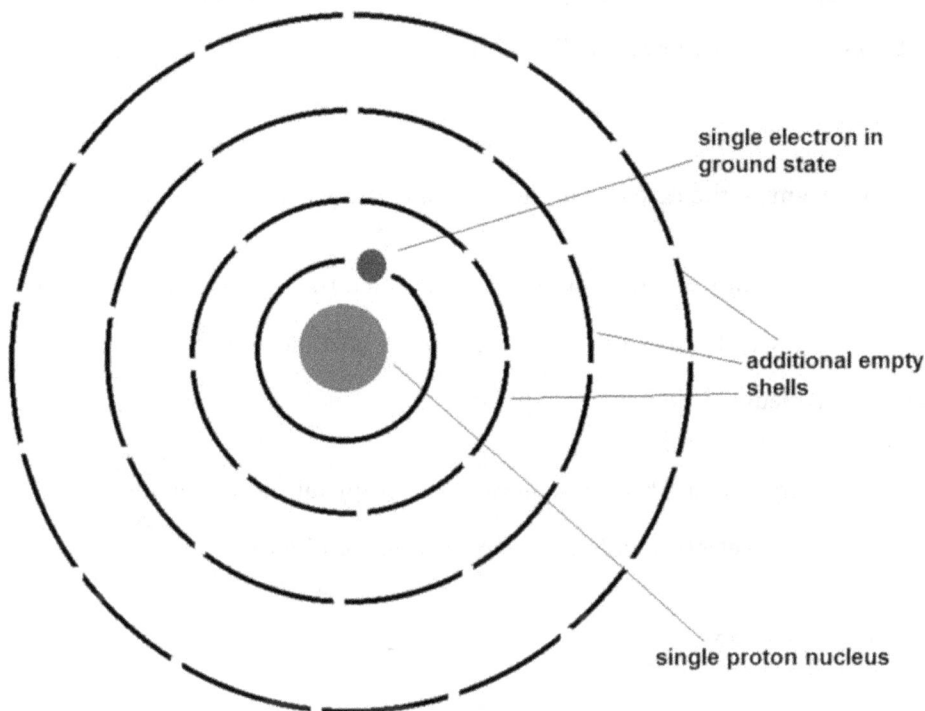

single electron in
ground state

additional empty
shells

single proton nucleus

Notice that there are a number of 'shells' that the electron can occupy.

Unless the atom is excited by the impact of energy from a number of possible sources, the atomic electron will remain in the shell closest to the nucleus which is called the 'ground state' corresponding to Principal Quantum Number 1.

Einstein realised that the atomic electron could be completely detached from the nucleus if sufficient energy impacted upon the atom.

Usually the energy is provided by an electromagnetic photon impinging upon the atom.

If the energy of the 'attacking' photon is insufficient to ionise the electron, the electron might be excited to a higher shell, from which it would collapse in a very short time indeed to a lower shell, and in so doing would emit an electromagnetic photon.

Let's now examine the case of ionisation of a hydrogen atom.

To ionise the atom means that the atomic electron must escape the nucleus altogether, in which case the electron would become a negative ion and the remaining nucleus would become a positive ion.

Now the energy of a photon required to completely ionise a hydrogen atom comes out at 13.6 electron volts, which is written as 13.6 eV.

What is an electron volt?

It is the amount of energy that is required to move an electron through a potential difference of 1V, and is given as some 1.6×10^{-19} Joules.

So if an Ultra Violet radiation photon of exactly 13.6 eV were to impinge upon a hydrogen atom, the atomic electron would become ionised – just!

It can be represented something like this:

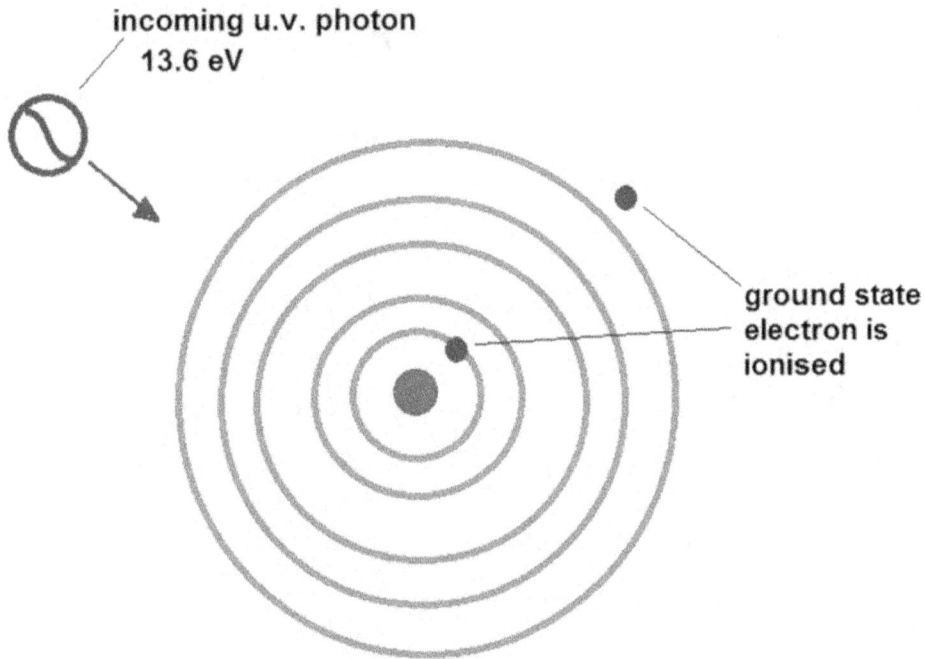

The problem is that quite quickly the electron would, in all likelihood, re-attach itself to the nucleus because it would not be moving.

However, if a photon of energy greater than 13.6 eV were to impinge upon the hydrogen atom, the 'photo-electron', which is the name given to the ionised electron, would have some kinetic energy.

The relationship between the energies is given by the following equation:

Energy of impinging photon = *work function energy of material + kinetic energy of photo-electron*

This can be written as:

$$E = \phi + E_K$$

This can be further reduced to:

$$hf = hf_0 + E_K$$

Where f_0 is the threshold frequency of the material being impinged upon – that is to say, it is the frequency of the impinging photon necessary for ionisation to occur.

So, for example, if a photon of energy 16 eV were to impinge upon a hydrogen atom, then the energy available for the kinetic energy of the photo-electron would be:

$$16 - 13.6 \ eV$$

This would allow some 2.4 eV to be available for the kinetic energy of the photo-electron.

Let us now calculate the exit speed to the photo-electron from the ionised nucleus.

$$E_K = \tfrac{1}{2} mv^2$$

Now the mass of an electron is approximately **9.11 x 10^{-31} kg** and the residual energy of the impinging photon must be given in joules.

Hence the residual energy of the impinging electron in joules is given as:

$$2.4 \ \times \ (1.6 \times 10^{-19}) \text{ joules}$$

Calculate it now and hence calculate the speed of the ionised electron.

What did you get?

The residual energy comes out at some **3.84 x 10^{-19} joules.**

This results in a speed of retreat of the photo-electron of some **(9.1816 x 10^{5}) ms^{-1}**

This is quite a surprising speed, but remember that the speed of an electron as it struck the now old fashioned T.V. tube is in the order of some one million metres per second!

So we have Albert Einstein to thank for the discovery that light was not a pure waveform, but could be viewed as discreet quanta (now called photons, of course).

His mathematical description of how the photoelectric effect was caused by absorption of light quanta appeared in his 1905 paper entitled "On a Heuristic Viewpoint Concerning the Production and Transformation of Light"

However, it must now be stated that the Rutherford model of the atom is very useful indeed in handling some of the energy equations in the quantum theory as we will see shortly.

Questions

1.

(a) Using the equation for the energy of a photon, which is:

$$E = h f$$

Where E is the energy of a photon, f is the frequency of the photon, and h is the Planck constant $(h = 6.6262 \times 10^{-34})$ Js, calculate the energy of a photon of frequency (1×10^{10}) Hz.

(b) Hence calculate the energy in electron volts (eV)

(c) Hence determine whether or not this photon can ionise atomic hydrogen

2. A photon of frequency (3.7×10^{15}) Hz impinges upon a hydrogen atom.

(a) Work out the energy of the photon and convert this into electron volts (eV)

(b) Will this ionise the electron?

(c) If so, calculate the speed of the photo-electron after ionization

3.   A potassium atom is impinged upon by a photon of ultra violet radiation of frequency $(1.2073285 \times 10^{15})$ Hz.

(a)   Calculate the energy of this photon in joules

(b)   What is this in electron volts (eV)?

(c)   If the threshold frequency of potassium is $(4.8293139 \times 10^{14})$ Hz, what is the required ionisation energy of a potassium electron? (Give your answer in joules).

(d)   What is this in electron volts (eV)?

(e)   Hence state whether or not the electron will be ionised when the ultra violet photon impinges on it

(f)   How many electron volts of energy are now available for the kinetic energy of the ionised electron, assuming it is ionised?

(g)   If the electron is ionised, calculate its speed.

Chapter 4:   Something About Light

Now, before Einstein's great discovery, it was thought that light was a continuous wave.

But it was Einstein who discovered that only electromagnetic radiation of high enough energy could ionise an atomic electron.

Had light been a continuous wave, then all one had to do was wait for a sufficient time to pass and the atomic electron would, presumably, have gathered sufficient energy to become ionised.

However, it didn't matter how long one waited, electromagnetic radiation of low energy would **never** ionise an atomic electron.

It was therefore Einstein who discovered that light, in fact, was quantised.

That is, that it existed in discreet packages called **photons**.

So in other words, Newton's idea of light being in the form of rays of particles is, in a sense, true.

Actually both ideas are true – light can be shown to be wavelike in nature.

Light can also be shown to be particulate in nature, the particles being photons of minimal energy.

This phenomenon whereby only light of sufficient energy can ionise an atomic electron is called the photoelectric effect.

It means that only one photon can impinge upon one atomic electron, and if the impinging photon is of insufficient energy to ionise the electron, it might excite the electron to a higher energy level temporarily.

The Bohr Atom

Niels Bohr, one of the great founding fathers of the quantum theory, postulated that a hydrogen atom was surrounded by *shells*.

These shells were always empty (in the case of a hydrogen atom), the sole atomic electron always occupying the ground state shell, which was nearest to the nucleus.

This can be represented thus:

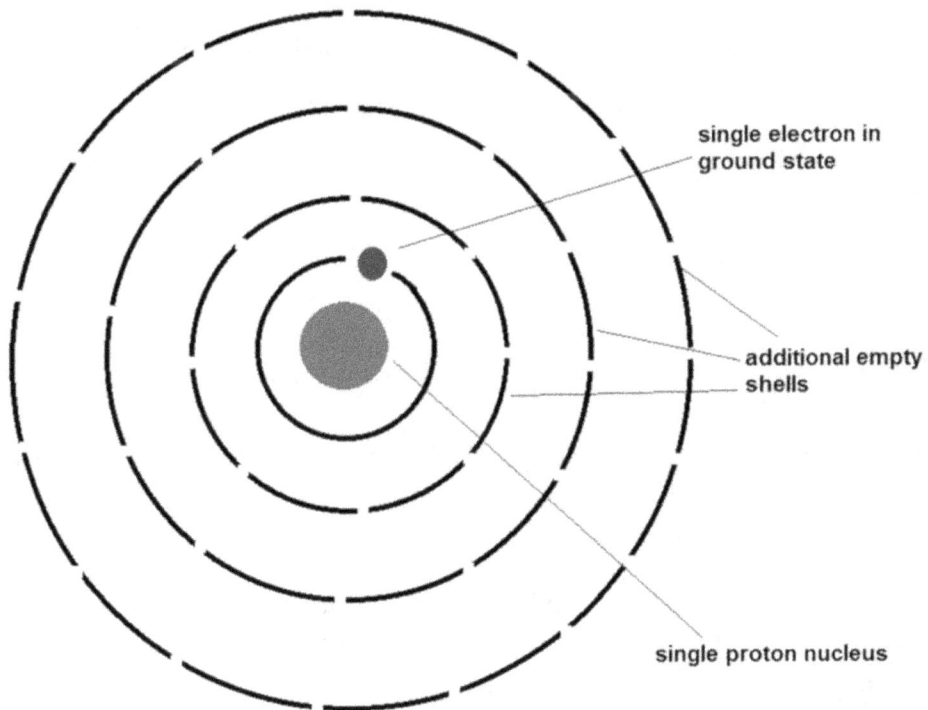

single electron in ground state

additional empty shells

single proton nucleus

Unless, of course, the atom was impinged by a photon of insufficient energy to ionise the electron.

In this case, the atomic electron might be excited into a higher energy level shell, in which case, after a very short time indeed (in the order of a hundred thousandth of a second), the electron would collapse again ending up in the ground state again.

If the atomic electron was excited into a very high energy level shell, it might not collapse immediately into the ground state, but instead it might go through the shells, and with each descent it would emit a photon of energy commensurate with the difference in energy levels between the shells.

It is this fact that enables microwave cookers to function

The atomic electrons in the food (usually concentrated in the moisture within the food), will excite atomic electrons to high shells, and the descent will be via a number of shells, with some collapses emitting infra red photons, thus giving heat to the food.

Count Louis de Broglie

The great quantum physicist Count Louis de Broglie became a Nobel Laureate in 1929.

His first degree was in history, and he then became an excellent physicist.

He understood that light quanta could be considered both waves and particles.

Since wavelike objects (like photons) could be considered particles, he proposed that maybe matter particles could be considered to be waves as well.

Initially, (as is all too often the way in physics), the establishment decried his ideas, just as the great Subrahmanyan Chandrasekhar's ideas, which were subsequently proved correct of course, about the mass limit of a white dwarf were decried initially.

However, de Broglie's ideas have been subsequently proved to be perfectly correct, and we now know that matter particles are wavelike in nature too, particularly the electron, which we will now consider.

De Broglie also produced an equation for working out the wavelength of a particle, as shown below:

$$\lambda = \frac{h}{mv}$$

where λ is the wavelength of the wave, **h** is the Planck constant, **m** is the mass of the particle, and **v** is its velocity, with **mv** being the momentum of the particle.

Another great founding father of the quantum theory, Werner Heisenberg, realised very quickly that 'solid' particles were wavelike in nature as well, one of his famous quotes being:

> **"We will have to abandon the philosophy of Democritus and the concept of elementary particles. We should accept instead the concept of elementary symmetries"** (1)

This concept was soon proved when the great G.P. Thomson decided to fire electrons through thin gold foil in 1927.

He succeeded in doing this and immediately discovered that the electrons exhibited an interference pattern, showing that they too were wavelike in nature!

Consequently, the model of the atom needed to be updated.

So along came the 'wave mechanics' model of the atom.

This does not show the electron as a point particle orbiting the nucleus, but instead it shows an array of possible probabilistic positions that the electron can occupy near the nucleus, with, of course, the greatest probability being in the ground state.

This can be represented thus:

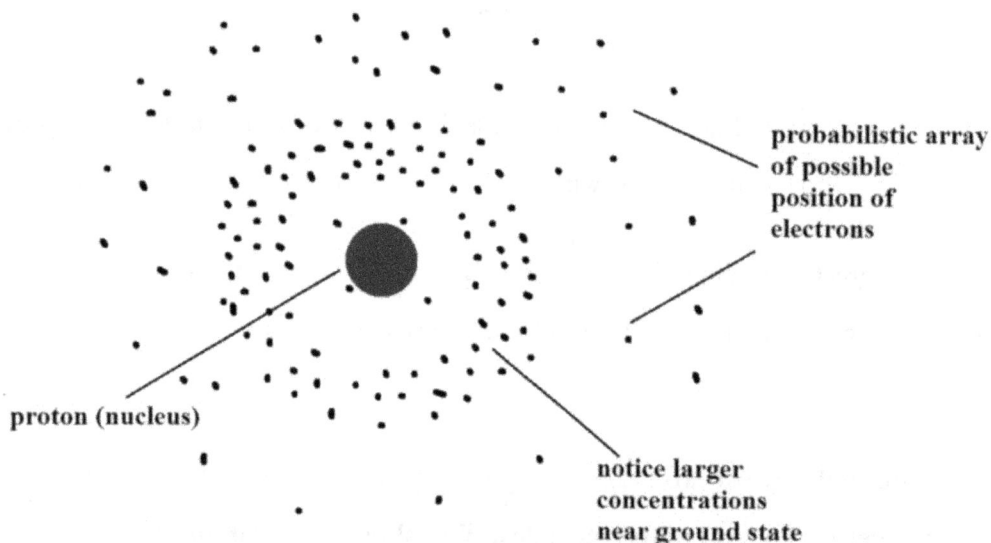

probabilistic array of possible position of electrons

proton (nucleus)

notice larger concentrations near ground state

However, remember that both the Rutherford model and the Bohr model of the atom are frequently used in physics, and correctly so.

Chapter 5:   Principal Quantum Numbers
and Bohr Radii

What are principal quantum numbers?

Principal quantum number, 'n', as applied to the shells surrounding a hydrogen atom are the numbers of the shells with the ground state given the principal quantum number of **1**.

The remaining principal quantum numbers increase as the shells go outwards.

This can be represented below:

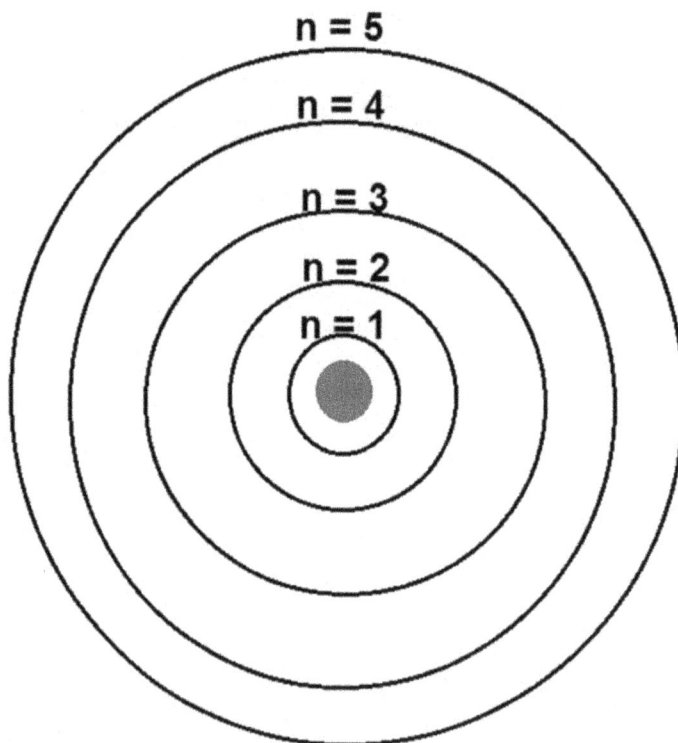

Can we actually work out the Bohr radii of these shells?

The answer is yes we can, so we will now derive an equation for this radius.

We will use the Rutherford model of the atom in which the atomic electron is assumed to be a point particle which orbits the single proton nucleus.

We first equate the two forces acting upon the atomic electron.

These are the centripetal force which acts in an inward direction which, in fact, is the inward pull of the electromagnetic force between the proton and the electron.

We now equate these:

$$\textbf{Centripetalforce} = \textbf{Electromagnetic force}$$

$$\frac{mv^2}{r} = \frac{e^2}{4\pi\,\varepsilon_0\,r^2}$$

Where, m = mass of electron, (electron mass $= 9.10956 \times 10^{-31}$ kg), r = Bohr radius, v = tangential speed of electron, e = electron/proton modular charge,(e $= 1.602192 \times 10^{-19}$ C), ε_0 = permittivity of free space($\varepsilon_0 = 8.85419 \times 10^{-12}$ F m^{-1})

Hence,
$$r = \frac{e^2}{4\pi\,\varepsilon_0\,mv^2}$$

Now, the electron will possess angular momentum, which is given as:

27

$$L_{ang} \quad = \quad \frac{mv}{r}$$

But we also know that angular momentum of the atomic electron is given as:

$$L_{ang} \quad = \quad \frac{nh}{2\pi}$$

Where, n = principle quantum number, and h = the Planck constant

$$(\ h = 6.6262 \times 10^{-34} \ J \ s \ )$$

Equating these two, we get the following value for 'v':

$$v \quad = \quad \frac{nh}{2\pi \ mr}$$

Inserting this into our original equation we get:

$$r_n \quad = \quad \frac{\varepsilon_0 \ n^2 \ h^2}{e^2 \ \pi \ m}$$

Where r_n is the Bohr radius for any shell.

Example

Calculate the Bohr radius for the shell at principal quantum number 3.

Solution

$$r_n = \frac{\varepsilon_0 \, n^2 \, h^2}{e^2 \, \pi \, m}$$

$$r_n = \frac{(8.85419 \times 10^{-12}) \times 3^2 \times (6.6262 \times 10^{-34})^2}{(1.602192 \times 10^{-19})^2 \pi \, (9.10956 \times 10^{-31})}$$

Hence,

$$r_n = 4.7626 \times 10^{-10} \text{ m}$$

Questions

1. Calculate the second and fifth Bohr radii for a hydrogen atom.

2. What is the size of the gap between the second and fourth Bohr radii for a hydrogen atom?

3. Calculate the radius of the ground state shell (principal quantum number 1) for hydrogen.

Chapter 6:   Energy Levels in Atomic Hydrogen

As energy is added to an atomic electron in hydrogen, the electron gains energy.

However, when the atomic electron becomes ionised and leaves the atom, its potential energy is actually **zero**.

Since the electron is gaining energy as it climbs the shells, how can it end up with a potential energy of zero?

In order to visualise this, let us now go to an analogy.

The analogy we will use is that of a being in a gravity field on a planet such as we now are.

Imagine if we could suddenly be transported into intergalactic space, away from all gravity fields and totally motionless with no speed at all.

What would our energy be?

It doesn't take much imagination to realise that our net energy would be zero.

So what happened to all the energy that was necessary to eject us from the Earth's gravity field (the escape velocity to get out of the Earth's gravity field is, by the way, is 11.2 km s^{-1}).

Imagine that in order to truly escape from the Earth's gravity field we need to achieve an upward speed of 11.2 km per second – and that means travelling through 11.2 km in each and every second.

But what is worse is the escape velocity to escape the Sun's gravity field, or otherwise we just be a permanent part of the Solar System.

The escape velocity for the Sun is 617.5 km s^{-1}.

In other words, unless we could achieve a speed in which we would cover 617.5 km in each and every second, we would never escape our Solar System.

What is even worse is the escape velocity needed in order to escape the gravity field of our galaxy, in order to reach intergalactic space.

The escape velocity needed for this is 525 km s^{-1}.

In other words, somehow we would need to cover 525 km in each and every second in order to get into intergalactic space.

This would take an enormous amount of energy, so how could our net energy lying motionless in intergalactic space be zero?

The answer is that we are marooned in a **negative energy well** here on earth, and in order to escape to intergalactic space, we would need to replace our negative energy with positive energy in order to achieve zero energy.

The same is true for a hydrogen atomic electron.

The electron is trapped in a **negative energy well** as it orbits the nucleus in the ground state, and it needs positive energy to be provided to neutralise the negative energy it is trapped in.

So the 13.55 eV of energy that must be provided by an ultra violet photon to ionise the electron will only bring that electron up to zero potential energy upon ionisation.

So although the atomic electron is gaining energy, it is simply losing negative energy on its way to the zero energy state of ionisation.

Can we actually calculate the amount of energy between the shells?

Yes we can and we will derive a simple formula to do that now.

Using the Rutherford model of the atom:

Total energy of atomic electron = kinetic energy of motion in orbit + potential energy

But the potential energy is negative, so:

Total energy of electron = kinetic energy of motion in orbit − potential energy

$$E_T = \tfrac{1}{2}mv^2 - F_E r_n$$

Now, $$F_E = \frac{e^2}{4\pi \varepsilon_0 r_n^2}$$

We also know that:

$$mvr = \frac{nh}{2\pi}$$

33

Hence:

$$v = \frac{n\,h}{2\pi\,mr_n}$$

We also know that:

$$r_n = \frac{\varepsilon_0\,n^2\,h^2}{e^2\,\pi\,m}$$

Hence:

$$E_T = \frac{-\,m\,e^4}{8\,\varepsilon_0{}^2\,n^2\,h^2}$$

Since this energy is negative, then energy must be added in order to raise the atomic electron to a higher shell.

We can now calculate the energy of an atomic hydrogen electron in any shell, by adding the appropriate principal quantum number, **n**.

Example

Calculate the energy of a photon necessary to ionise a hydrogen atomic electron from the ground state.

Since the electron is in the ground state, with principal quantum number 1, we must use this principal quantum number in our equation in order to calculate the photon energy required to ionise the atomic electron.

$$E_T = \frac{-m\,e^4}{8\,\varepsilon_0{}^2\,n^2\,h^2}$$

Using the data previously given, we have:

$$E_T = \frac{-m\,e^4}{8\,\varepsilon_0{}^2\,n^2\,h^2}$$

$$E_T = \frac{-\,(9.10956 \times 10^{-31}) \times (1.602192 \times 10^{-19})^4}{8\,(8.85419 \times 10^{-12})^2 \times 1^2 \times (6.6262 \times 10^{-34})^2}$$

$$E_T = 2.1799117 \times 10^{-18} \text{ joules}$$

Converting this to electron volts (eV), we have:

$$E_T = \frac{2.1799117 \times 10^{-18}}{1.602192 \times 10^{-19}}$$

$$E_T \;\; = \;\; 13.6 \text{ eV (approximately)}$$

Questions

1. Calculate the energy required to excite a hydrogen atomic electron from the ground state to principal quantum number 2. Give your answer in both joules and electron volts.

2. Calculate the energy of the photon emitted when a hydrogen atomic electron descends from the shell at principal quantum number 5 to the shell at principal quantum number 4. Give your answer in both joules and electron volts.

3. Calculate the energy necessary of an impinging photon on a hydrogen atomic electron to cause it to ascend from the ground state shell, of principal quantum number 1 to the shell at principal quantum number 3. Give your answer in both joules and electron volts.

Chapter 7:   The Rydberg Constant

Johannes Rydberg (1854 – 1919) was the physicist who first showed us how to calculate the actual wavelength of any photon which was emitted from a hydrogen atom when the atomic electron descended from a higher to a lower shell.

We now derive a simple equation to enable us to calculate this:

The expression for the energy level in any shell is given as:

$$E_T \;=\; \frac{-\,m\,e^4}{8\,\varepsilon_0^{\,2}\,n^2\,h^2}$$

Hence the difference in energy levels between any two shells given as shell 1 and shell 2 is as follows:

$$E_2 \,-\, E_1 \;=\; \frac{(-\,m\,e^4)}{(8\,\varepsilon_0^{\,2}\,n_2^{\,2}\,h^2)} \;-\; \frac{(-\,m\,e^4)}{(8\,\varepsilon_0^{\,2}\,n_1^{\,2}\,h^2)}$$

$$E_2 \,-\, E_1 \;=\; \frac{-m\,e^4}{8\,\varepsilon_0^{\,2}\,h^2}\; \frac{[1]\,-\,[1]}{[n_1^{\,2}]\,-\,[n_2^{\,2}]}$$

But

$$E \;=\; h\,f$$

And

$$c = f\lambda$$

Hence,

$$E = \frac{hc}{\lambda}$$

Hence the wavelength emitted when a hydrogen atomic electron descends to a lower principal quantum number shell is given as:

$$\frac{1}{\lambda} = \frac{m\,e^4}{8\epsilon_0^{2}h^3} \times \frac{[1}{c[n^2_1} - \frac{1}{n^2_2]}]$$

Remembering now that the following data apply:

Permittivity of free space,

$$\epsilon_0 = 8.8542 \times 10^{-12} \text{Fm}^{-1}$$

Planck's constant,

$$h = 6.6263 \times 10^{-34} \text{Js}$$

Speed of light in vacuo,

$$c = 2.998 \times 10^{8} \text{ms}^{-1}$$

We can now calculate the energy difference in descention between shells

Notice that the multiplier of the Principal Quantum Number difference is a constant.

This is called the **Rydberg Constant**.

Calculate the value of this constant now, and then look up the answer in the last chapter.

Hence the Rydberg equation now becomes:

$$\frac{1}{\lambda} = (1.097 \times 10^7) \times \left[\frac{1}{n_1^2} - \frac{1}{n_2^2}\right]$$

Notice that since this constant is formulated from some other invariant constants of the Universe, we can regard this constant as one of the basic physical constants of the Universe, where λ is the wavelength of the radiation emitted in vacuo.

Questions

In the following examples, use the data given below:

Type of radiation	Wavelength (m)
Microwaves	(1×10^{-2}) to 10
Infra-red	(1×10^{-5}) to (3×10^{-7})
Visible light	(4×10^{-7}) to (7×10^{-7})
Ultra violet	(8×10^{-7}) to (5×10^{-8})

1. Calculate the wavelength of the photon emitted when a hydrogen atomic electron collapses from the shell with principal quantum number 3 down to 2.

   What type of radiation is this?

2. Calculate the wavelength of the photon emitted when a hydrogen atomic electron collapses from the shell with principal quantum number 4 down to 1.

   What type of radiation is this?

Chapter 8:   The Schrödinger Equation

The great Austrian physicist Erwin Schrödinger was one of the very important founding fathers of the quantum physics and he succeeded Max Planck to the chair of theoretical physics in Berlin.

He was greatly influenced by Count Louis de Broglie and he also correctly took the view that a wave is always associated with each free particle.

His equation has been proved to be extremely reliable and he was awarded the Nobel Prize along with Paul Dirac in 1933 for their important contribution to "New and productive forms of atomic theory".

One form of the time independent Schrödinger equation is given below:

$$\overset{\textbf{wave function, } \Psi}{\frac{d\Psi}{dx^2} + \frac{2m}{\hbar^2} \left(E_{tot} - E_{pot}(x) \right) \Psi = 0}$$

In some ways, the most important feature of the Schrödinger equation is the wave function, Ψ.

The square of this function will give us the probability of where an unobserved quantum particle might be in spacetime.

Quantum Tunnelling

It is well known and documented that unobserved quantum particles can be in multiple places at once.

What is more counterintuitive is that quantum particles can be where they shouldn't be at all.

For example, suppose we have an electron positioned near a negative barrier such that the negative potential, **V,** on the barrier is such that the electron cannot penetrate the barrier.

In actual fact, there is a distinct probability that the electron can and, in fact, must penetrate the barrier, like this:

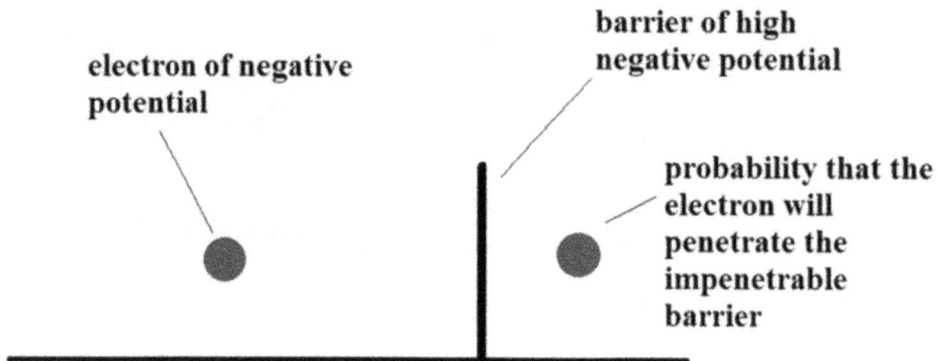

The only real explanation for this phenomenon is that, somehow, the electron was already there!

Quantum tunnelling is a well observed phenomenon, and were it not for this phenomenon, we wouldn't have transistors, nor would we be able to smell, since the sense of smell is also dependent on quantum tunnelling.

We will now perform a simple calculation using the time independent version of the Schrödinger equation, which will serve to demonstrate the probabilistic nature of quantum tunnelling.

Example:

A particle of mass **m = (2.0 x 10^{-29}) kg** and total energy of **3.0 eV** possesses the potential energy function shown below, which consists of a square potential energy barrier of at **5.0 eV** and width of **(2.5 x 10^{-10}) m**:

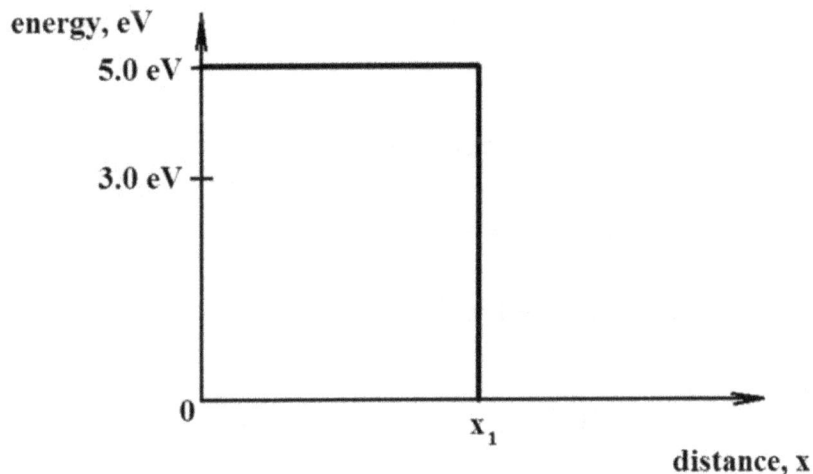

The wavefunction which describes the particle can be written as follows:

$$\Psi(x) = A\,e^{-kx}$$

A and **k** being constants.

If the probability of encountering the particle near **x = 0** is **0.3**, then what is the probability of finding the particle in a similar region near $x = x_1$?

Solution

In line with our equation, we need to differentiate the given wave function twice:

$$\frac{d\Psi}{dx} = -A\,k\,e^{-kx}$$

$$\frac{d^2\Psi}{dx^2} = A\,k^2\,e^{-kx}$$

Substituting in our main equation:

$$A\,k^2\,e^{-kx} - \frac{2m}{\hbar^2}(E_{tot} - E_{pot})\,x\,A\,e^{-kx} = 0$$

Now since e^{-kx} cancels out on both sides, we know that:

$$k = \frac{\sqrt{(2 \, m \, (E_{tot} - E_{pot}))}}{\hbar}$$

$$k = \frac{\sqrt{(2 \, (2.0 \times 10^{-29}) \, (8.0 \times 10^{-19} - 4.8 \times 10^{-19}))}}{(1.06 \times 10^{-34})}$$

So that:

$$k = 3.375 \times 10^{10} \, m^{-1}$$

Having converted electron volts into joules

Now the ratio of probability of finding the particle near $x = x_1$ is given as:

$$\frac{p_2}{p_1} = \frac{(A \, e^{-kx}_1 \Delta x)}{(A \, e^{-k0} \Delta x)}$$

$$\frac{p_2}{p_1} = \frac{(\exp(-3.375 \times 10^{10} \times 1.2075 \times 10^{-10}))}{1}$$

Since $$e^0 = 1$$

$$\frac{p_2}{p_1} = 0.0002886$$

Now, since $$p_1 = 0.3, \quad \text{then:}$$

$$p_2 = 8.6566 \times 10^{-5}$$

Question

A particle of mass (1.6×10^{-29}) kg and total energy of 2.3 eV has the energy function shown below.

This consists of a square wave barrier of height 4.4 eV and width of (3.0×10^{-10}) m.

This is shown in the following diagram:

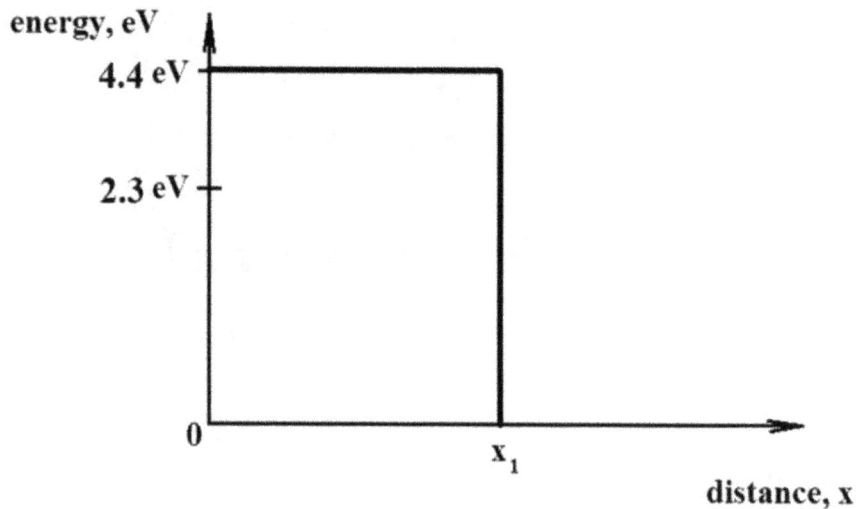

(a) Calculate the value of the constant **k** in the following equation:

$$\frac{d\Psi}{dx} = -A\,k\,e^{-kx}$$

(b) If the probability of finding the particle in a region near $x = 0$ is **0.2**, calculate the probability of finding the particle in a similar region near $x = (3.0 \times 10^{-10})$ **m**.

Now this probability is really the probability of the process taking place which is known as **Quantum Tunnelling**, in which case a quantum particle can be seen (probabilistically speaking) on the other side of a barrier.

It is this effect that enables transistors and solid state diodes to work, since the quantum particles must penetrate a barrier in these cases.

How, in the above example, could we increase the probability of the tunnelling taking place?

In the above example, if kx_1 was smaller, there would be a greater probability of the tunnelling taking place.

Chapter 9:   The Wave Function, Ψ

We will be very concerned in the next few chapters about the wave function and its nature.

Remember that the square of the wave function, Ψ, will give us the probability as to where a quantum particle will be at any instance.

The problem is that if a quantum particle is observed, the wave function is said to 'collapse' according to the Copenhagen Interpretation of the quantum theory.

These meetings, in the 1920s were of some of the founding fathers of the quantum theory, namely Neils Bohr, Werner Heisenberg, Paul Ehrenfest, Lise Meitner, Paul Dirac, Max Delbruck and others.

The general outcome and conclusion of these meetings was that physicists should use the equations of the quantum theory (referred to colloquially as 'quantum cookery') but give no thought to the philosophical arguments as to what exactly is going on in the quantum world.

Now, it must be said, that the equations of quantum theory are highly successful, and they point to one of the most successful theories in physics ever produced.

However, the reasons as to why the quantum theory works and exactly what is going on seems to be very counterintuitive.

We will examine in some detail exactly what we think is going on here.

It must be said that many quantum theorists did disagree with the consequences of the Copenhagen meetings, not least of whom was Albert Einstein, who then wasted decades in trying to disprove the weirdness of the quantum theory, along with, notably, Boris Podolsky and Nathan Rosen.

However, when John Bell came along, although unfortunately he died before his inequality could be shown to be correct, the famous EPR thought experiment which was meant to show the ideas behind the quantum theory were wrong, his inequality was proved, eventually to be completely correct, with the first experiment in this attributed to the physicist Alain Aspect in France.

Chapter 10:   The Laser

We know that an excited atomic electron which is temporarily occupying a shell above the ground state, will collapse in something under one hundred thousandth of a second.

However, in 1917, Albert Einstein showed us a variation in this process.

In fact he showed us that there is another way in which an excited atomic electron can be made to collapse.

This is in fact **Stimulated Emission**.

And this is how it operates:

Imagine an impinging photon on to an atom which has an excited atomic electron.

The impinging photon must have exactly the same energy as the photon that the atomic electron would emit upon spontaneous collapse.

If the impinging photon, of exactly the same energy, impinges upon the atom with the excited atomic electron, then the atomic electron is made to undergo a collapse with the identical energy of the impinging photon.

In other words, if the energy that the photon emitted from the collapse of the atomic electron was to be say **5.3 eV**, then the impinging photon must have exactly an energy of **5.3 eV**.

If this happens, then the emitted photon from the atomic electron collapse would be **in phase** with the impinging photon, would be emitted in the **identical direction** as the impinging photon, have **exactly the same energy** as the impinging photon, and could be said to be **coherent** with the impinging photon.

This now means that there would be two identical photons travelling in the same direction.

Suppose now that these two photons release another two identical photons from excited atomic electrons from two more atoms that they encounter.

What, in fact, would we have?

The answer is that we would have **amplification**.

And this is, pictorially, what we would have:

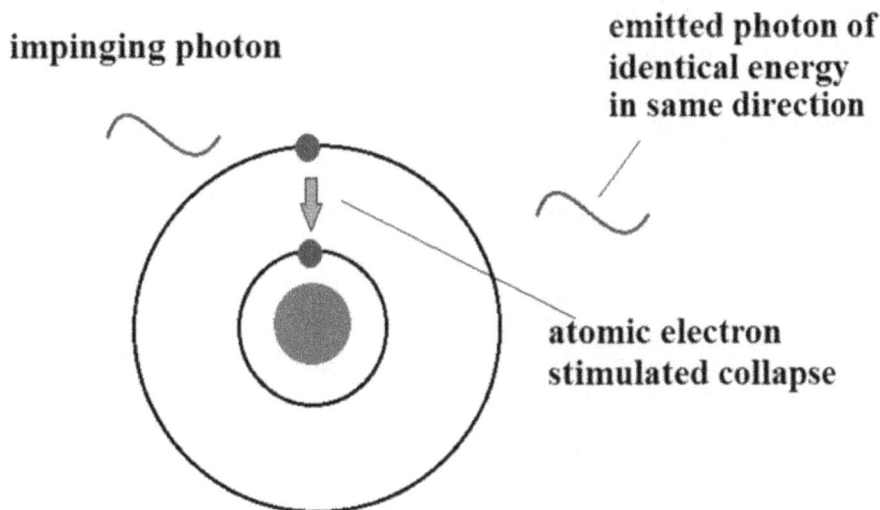

impinging photon

emitted photon of identical energy in same direction

atomic electron stimulated collapse

Or this:

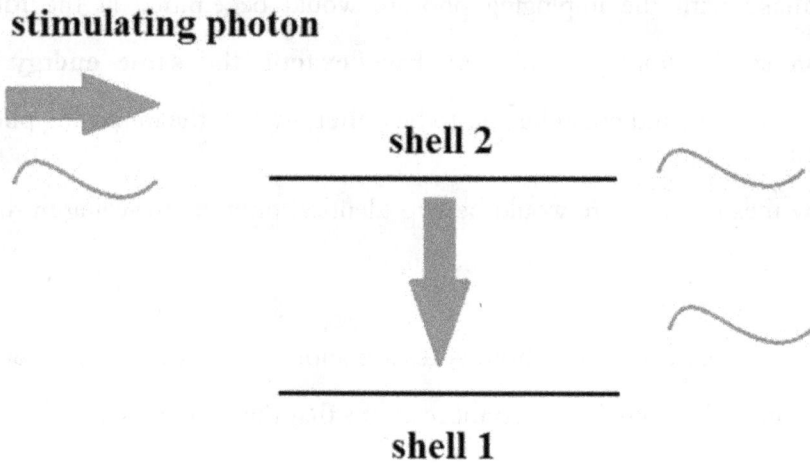

stimulating photon

shell 2

shell 1

And if the two (now impinging) photons impinged upon another two atoms with atomic electrons excited at identical amounts, we would have a total of four photons ready to impinge upon another four atoms with atomic electrons excited at the right amount.

This is amplification, like this:

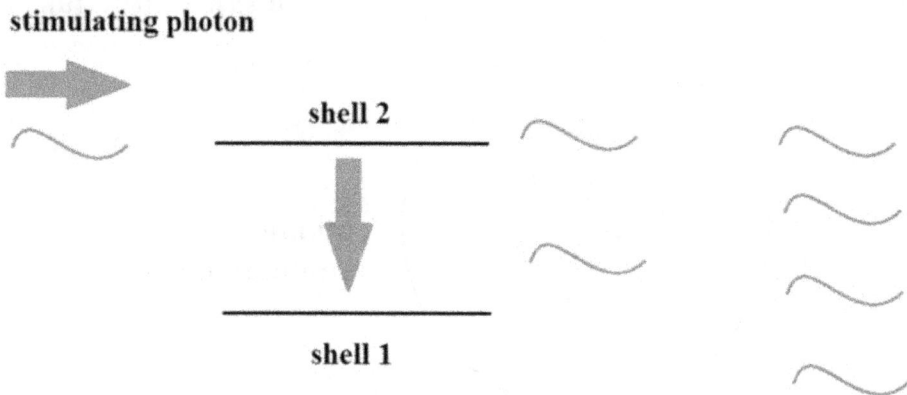

stimulating photon

shell 2

shell 1

In theory, this would now give us some sixteen impinging photons which, if they fell upon another four atoms with atomic electrons excited to the right amount, would give us eight impinging photons.

This type of amplification is referred to as **optical amplification**.

This is the nature of stimulated emission, and in fact, the term **laser** stands for **'light amplification by stimulated emission of radiation'**.

And, basically, this is how laser beams are generated.

In 1958, physicists Charles Townes and Arthur Schawlow describe how this could be achieved in practice.

And in the 1960s, the first laser was built.

In order for a laser to work, we must first have an **Active Medium**.

This can be any material that can be made to emit light photons, one example being **synthetic ruby**.

In what state will the atomic electrons be in any active medium that is unexcited?

The atomic electrons in such a case will be in their ground state.

In the diagram below, the active medium is shown and note that all the atomic electrons are in their ground state.

Notice also that at one end of the laser tube, there is a fully reflective mirror, whilst at the other end there is a partially reflective mirror.

Notice also that the partially reflective mirror will only allow some 1% to 5% of the photons hitting it to pass through it.

In other words, some 95% to 99% of the photons will be reflected back into the active medium when the medium is lasing:

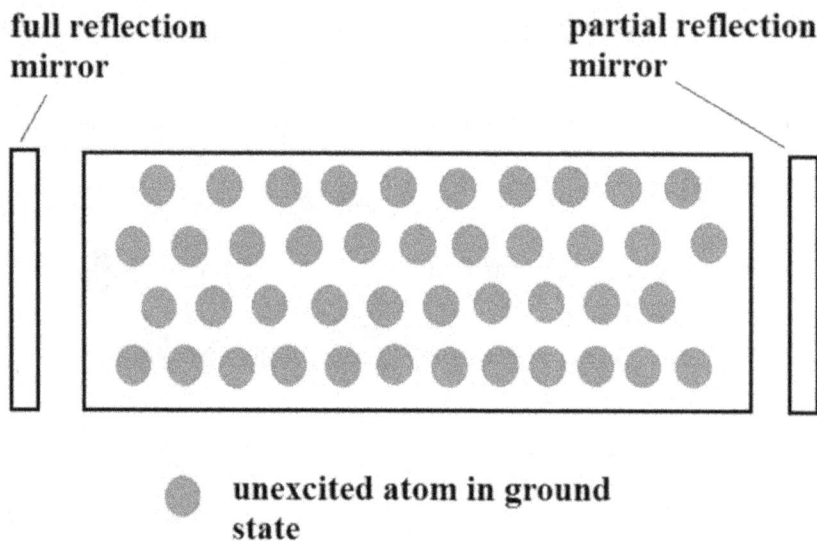

full reflection mirror

partial reflection mirror

 unexcited atom in ground state

Now if just a few of the atomic electrons are in an excited state, as shown in the following diagram, then the population of atoms is said to be **non inverted**, meaning that the active medium will **not** lase.

This is shown below:

full reflection mirror

partial reflection mirror

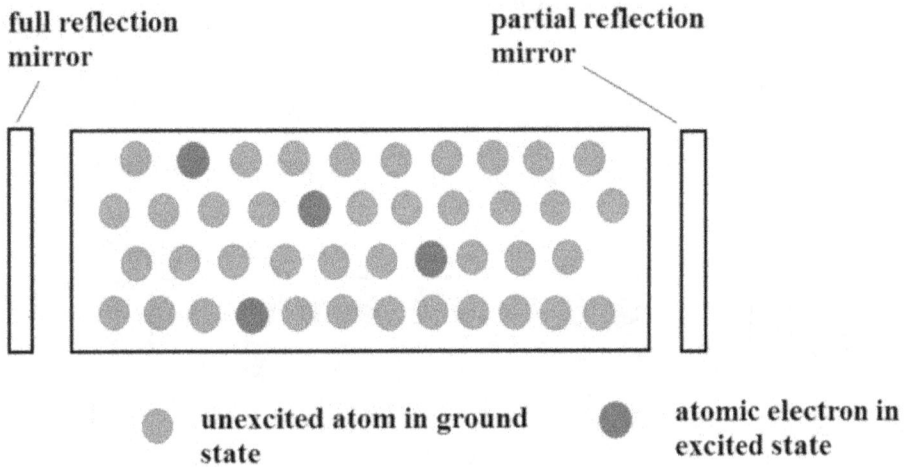

| | unexcited atom in ground state | | atomic electron in excited state |

However, if we can achieve a state where the majority of the atoms have atomic electrons in an excited state, then we have an active medium with an **inverted** population of atoms.

This is shown below:

full reflection mirror

partial reflection mirror

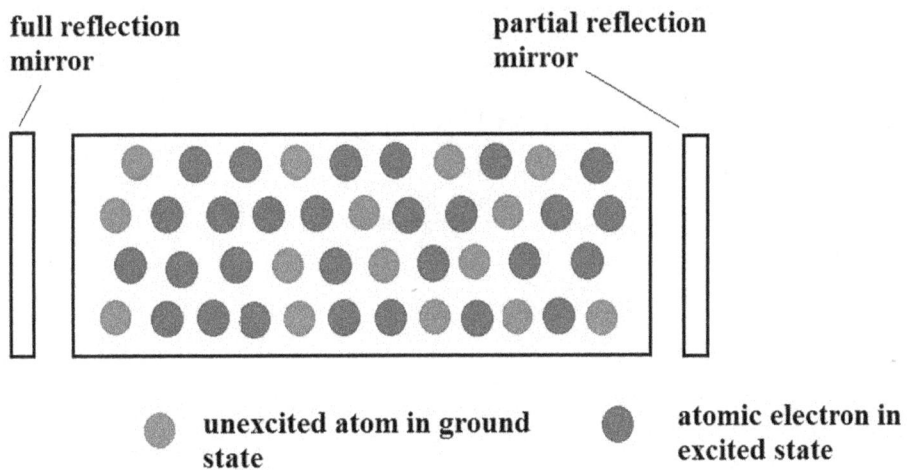

| | unexcited atom in ground state | | atomic electron in excited state |

However, what we need to achieve is a population in which **all** the atoms are in an excited state, giving us **full inversion**.

This is shown thus:

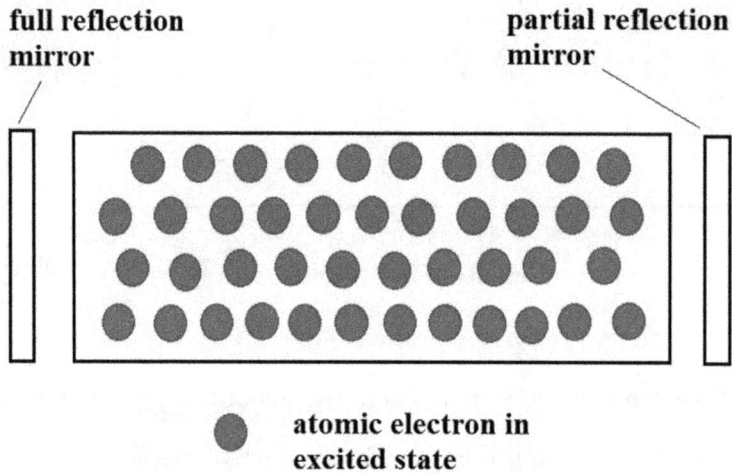

full reflection mirror

partial reflection mirror

atomic electron in excited state

How can we achieve this?

It is achieved when one of the atomic electrons undergoes **spontaneous emission**.

But we do not know the direction in which its photon, upon release, will go.

Maybe it will be in the direction shown below:

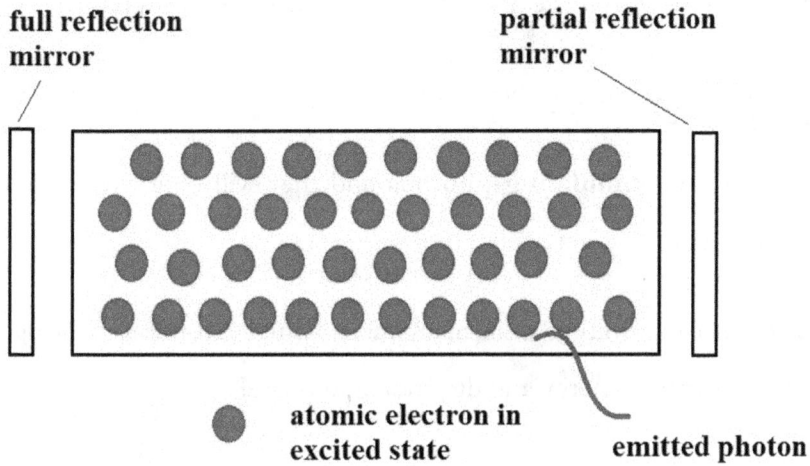

full reflection mirror   **partial reflection mirror**

● atomic electron in excited state **emitted photon**

In this case, we lose the photon since it isn't directed towards the mirrors.

However, we will soon have many photons emitted which are directed towards the mirrors, like this:

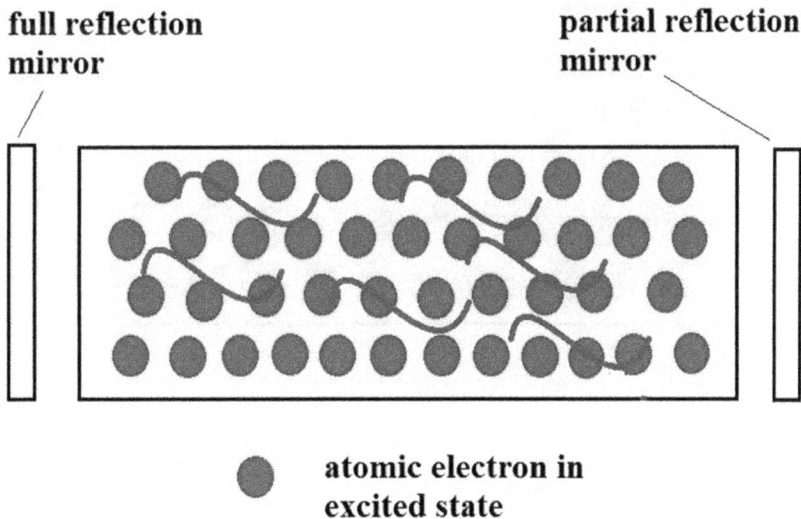

full reflection mirror   **partial reflection mirror**

● atomic electron in excited state

This now means that the photons will be fully reflected off the full reflection mirror and 95% to 99% of them will be reflected off the partial reflection mirror.

What have these photons now become?

They have become **stimulating photons** and they will continue to be reflected back and forth along the laser axis.

Can you see that this will result in many many stimulated photons being emitted, all of them of identical energy and direction and phase?

In order to achieve a good supply of stimulating photons, we need the operation of **optical pumping**.

To achieve this we need a high intensity flash lamp situated along the laser axis like this:

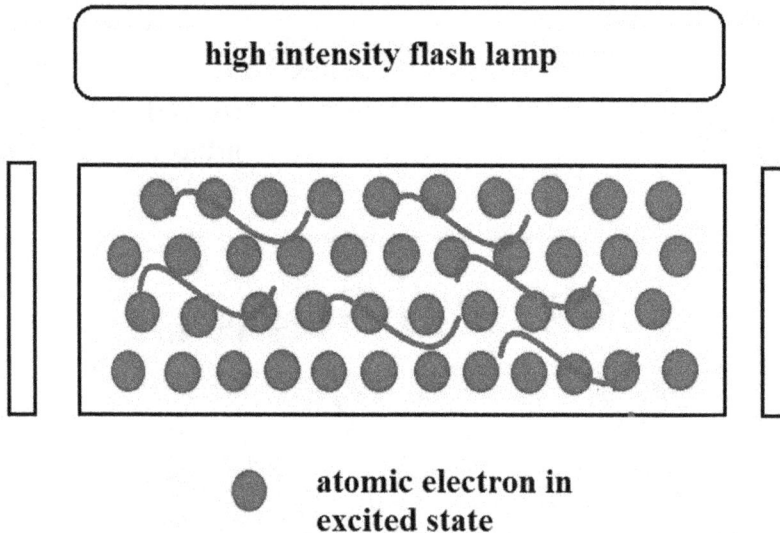

atomic electron in excited state

In this case, the photons from the flash lamp will stimulate the atomic electrons in the active medium.

Alternatively, the flash lamp can be 'wound around' the laser tube like this:

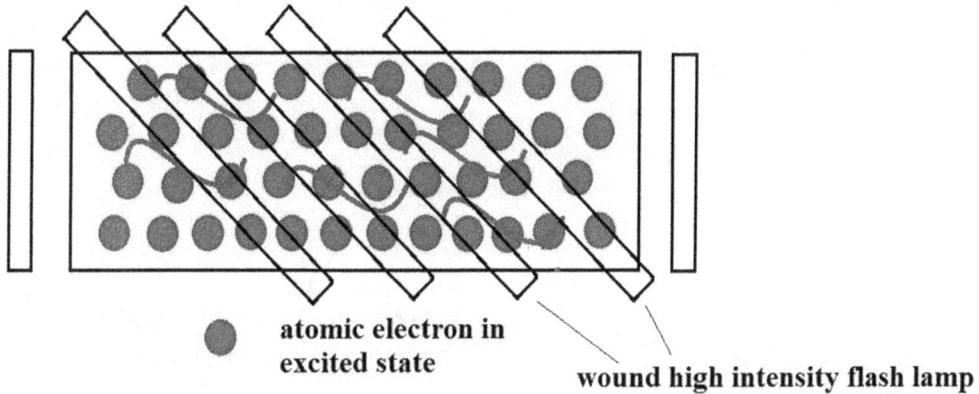

atomic electron in excited state

wound high intensity flash lamp

The Ruby Laser

The active medium here is **synthetic ruby**. This is **corundum**, which is Al_2O_3.

However, roughly one out of every one thousand aluminium ions is replaced by a **chromium ion**.

The photons from the high intensity flash lamp will result in the chromium ions Cr^{3+} achieving an excited **nuclear** state (as opposed to an excited atomic electron state).

Consequently, the relaxation will result in **no photon** being emitted by the atomic electrons, since it is the nucleus that is relaxing.

This relaxation will result in a **metastable state** being achieved, and relaxation from the metastable state will result in shell transfer occurring with the atomic electrons, and photon emission will result.

This can be shown on a quantum energy level diagram, and the one below shows excitation to **two excited nuclear states**, followed by atomic electron collapse with red light photon emission:

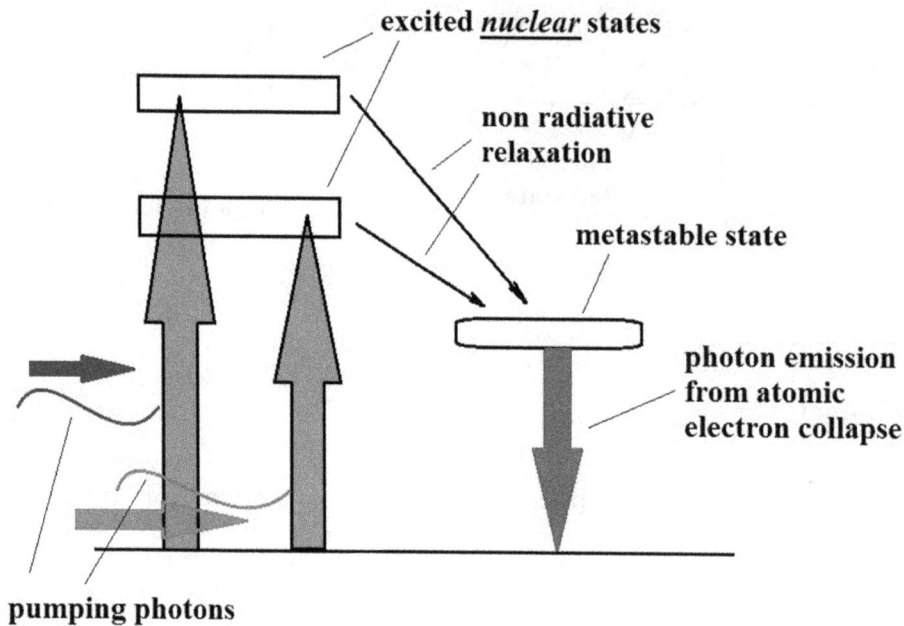

excited *nuclear* states

non radiative relaxation

metastable state

photon emission from atomic electron collapse

pumping photons

The chromium ions absorb photons in the green and blue light regions and the eventual red photon emission is of wavelength **694.3 nm**.

Remember that it is only the photons that are emitted along the axis of the laser that we are interested in – any photons in other directions will be lost.

Stimulation is achieved by **optical resonance**.

This is simply the fully reflecting mirror at one end and the 95% to 99% reflecting mirror at the other end.

Remember that since such a large quantity of photons will be reflected back and forth along the laser axis, a huge stimulated emission will result.

Any light which does pass through the partially reflecting mirror will be a highly collimated coherent beam of laser light.

This is shown below:

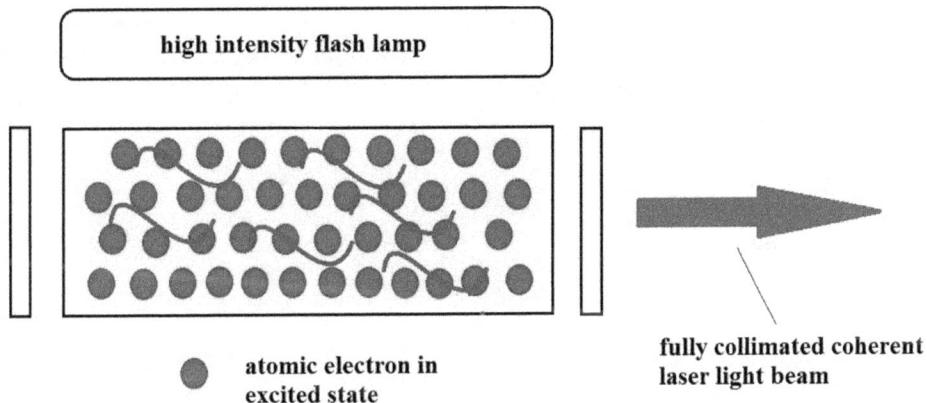

In summary, spontaneous emission is produced by the high intensity flash lamp.

Any spontaneously emitted photons that are not along the laser axis will be lost, but there will be many that will be directly along the axis.

The vast majority of these will be reflected back and forth along the laser axis, giving rise to a very large cohort of stimulated atoms, the atomic electrons of which will all be excited to exactly the same energy levels.

The small number of photons escaping from the partially reflecting mirror will be in the form of a highly collimated and coherent laser light beam.

Besides the ruby laser, there are many other types of active media.

Some of these include the following: carbon dioxide laser, sapphire mode locked laser, argon ion laser, helium neon laser, YAG laser (Yttrium Aluminium Garnet laser – neodymium doped).

Chapter 11:   Uses of Lasers

Many of the uses of the laser are well known.

These uses include the following: supermarket bar codes, transmission of information using fibre optics, laser eye surgery, the use of CDs, DVDs and Blu Ray discs.

Also, high power lasers can be used to cut metal.

However, a very important possible future use of lasers is to produce **nuclear fusion**.

Nuclear power stations at the time of writing use **nuclear fission**, and this is not to be confused with nuclear fusion.

Nuclear fusion is the process whereby energy is produced in stars, like our own Sun.

In the National Ignition Facility in Livermore, California, experiments are at an advanced stage to try to created nuclear fusion using lasers.

The process is as follows: a pellet, which will be made from either beryllium, plastic, or high density carbon, and which will contain both deuterium (one proton and one neutron) and tritium (one proton and two neutrons).

This will be bombarded with high intensity laser light from a number of laser sources, which will greatly increase the pressure on the contents.

The resulting pressure is sufficient to increase the density of the contents to some 20 to 100 times the density of lead.

The compression will only last for one trillionth of a second, but it should be sufficient to heat the deuterium and tritium to a temperature of around a thousand degrees kelvin.

This temperature will result in the pellet material exploding, thus raising the temperature of the deuterium and tritium to such a level that nuclear fusion should take place.

How big is the pellet?

Only about 2mm in diameter, like this:

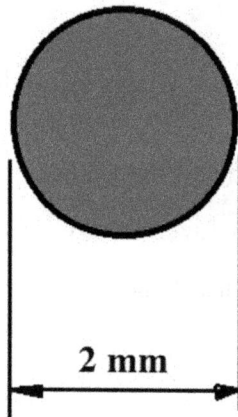

pellet containing deuterium and tritium

If this process were to be repeated some ten times per second in a power plant, we could have nuclear fusion as opposed to nuclear fission.

There are no radioactive products with nuclear fusion, apart from the surrounding material that is designed to absorb the neutrons that are released from the process.

Whatever is continually bombarded with neutrons over a period of time will become neutron heavy and will become unstable and radioactive.

The equation for the process is as follows:

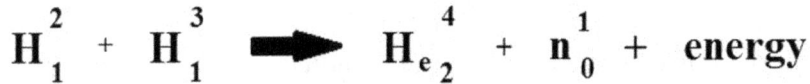

$$H_1^2 + H_1^3 \rightarrow He_2^4 + n_0^1 + \text{energy}$$

Note that the energy gained is **kinetic energy**, as with all nuclear processes, and this energy is converted in the surrounding material which serves as a means to turn the kinetic energy into **infra red radiation**, as the neutrons bounce off the atoms in the material, thus releasing infra red photons.

The amount of energy released in this case will be **17.59 MeV** per neutron.

Chapter 12:   The Nature of Light

In the late sixteen hundreds there was some controversy about the nature of light.

Isaac Newton thought that light was a ray or a stream of particles, but Christiaan Huygens believed it to be a wave.

But because of Newton's status, everybody went along with his view.

And then, of course, in the early eighteen hundreds, Thomas Young proved that light was wavelike with his famous double slit experiment, in which he was able to show that light actually formed an **interference pattern**.

Now waves can exhibit **four** basic properties.

These are **reflection, refraction, diffraction and interference.**

The problem is that not only waves will obey the laws of reflection.

Snooker balls will also obey the laws of reflection, and will bounce off a cushion with the angle of reflection equal to the angle of incidence.

Refraction is the phenomenon whereby a stick held under water appears bent.

However, it can be shown also that in certain circumstances, matter objects can behave in the same way.

Diffraction is the bending of a wave around a corner.

But under certain circumstances, material objects can display the same phenomenon.

The only attribute of waves that **only waves exhibit** is the phenomenon of interference.

This phenomenon is easily recognised.

The phenomenon is observed by a double sized wave appearing where two waves of equal amplitude meet, crest to crest and trough to trough, and where a trough meets a crest the wave disappears altogether and an area of calm forms.

This can be represented as follows:

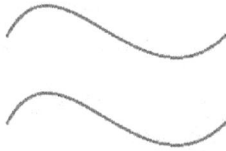

two waves in phase   **will result in a double amplitude wave**

And if the waves are in antiphase, this will result (as long as the waves are of equal amplitude:

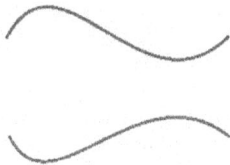

two waves in antiphase   **result in an area of calm with wave destruction**

The phenomenon of a double sized wave is called **constructive interference** or **reinforcement**.

The phenomenon of two wave meeting in antiphase is called **destructive interference**.

In a continuous display of interference, the regions in which constructive interference occurs are called **antinodal lines**.

However, if the original waves are not of the same amplitude, then the resulting waveform will be a wave of greatly reduced amplitude.

If the two waves are passed through a double slit or two sets of circular waves are produced by other means, then they will interfere.

At areas where crests meet troughs and destructive interference takes place, we will have a *nodal line*.

However, where two crests meet (and consequently two troughs), we will have **antinodal lines**.

The nodal lines are areas of complete calm (if the amplitude of the two original waves are identical) and the antinodal lines are where reinforcement takes place and will exhibit areas of double sizes waves (assuming the amplitude of the original waves are identical).

And the regions in which destructive interference occurs are called **nodal lines**, as is shown below:

In the above diagram, the antinodal lines are shown as black and white wavefronts, and the nodal regions are shown as completely black regions.

Chapter 13:   Back to Newton's Ideas

However, although photons can display wavelike properties, they can also display particle like properties.

For example, high energy electromagnetic waves possess **momentum**.

Arthur Compton (1892 – 1962) won the Nobel Prize in 1927 for what is known as **The Compton Effect**.

This physicist proved that not only did high energy radiation like X-rays lose energy when they collided with an electron (observed by an increase in the X-ray wavelength), but he observed that the X-ray photon would leave the electron at different angles, which, together with the angle at which the electron moved off, gave indication that the X-ray possessed **momentum**.

This could only mean that somehow, the X-ray photon possessed mass in some form.

We now know that all photons possess relativistic mass, although their rest mass is always **zero**.

In 1924, when Count Louis de Broglie was a graduate student in the University of Paris, he suggested that since photons could, under certain circumstances, decide to act as particles, he postulated that maybe solid particles (such as electrons) might be wavelike in nature too.

So in 1927, G.P. Thomson, who was the son of J.J. Thomson, who discovered the electron, decided to put de Broglie's idea to the test.

He decided to fire electrons through thin gold foil and to see what the result was.

Why did he choose thin gold foil?

Because the spacing between the atoms in gold was small enough to match the wavelength of the electron, should it be a wave.

His result was to change the face of physics for ever.

The results actually gave an interference pattern, which proved beyond doubt that electrons were also wavelike in nature, under certain circumstances.

Electrons are now fired through a carbon lattice in the modern version of this experiment, shown like this:

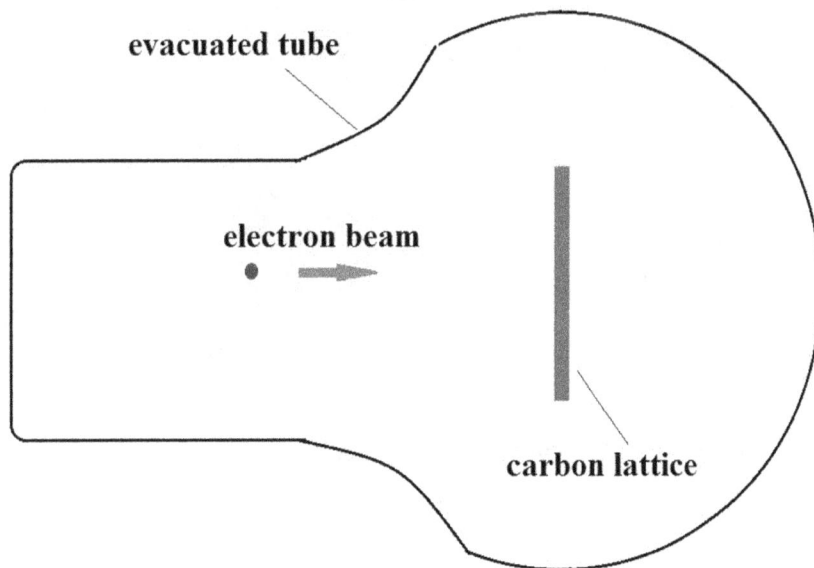

The interference pattern is displayed on the curved face of the tube, which is internally coated with a fluorescent material such as zinc sulphide.

The interference pattern obtained, looking at the front of the evacuated tube, is something like this:

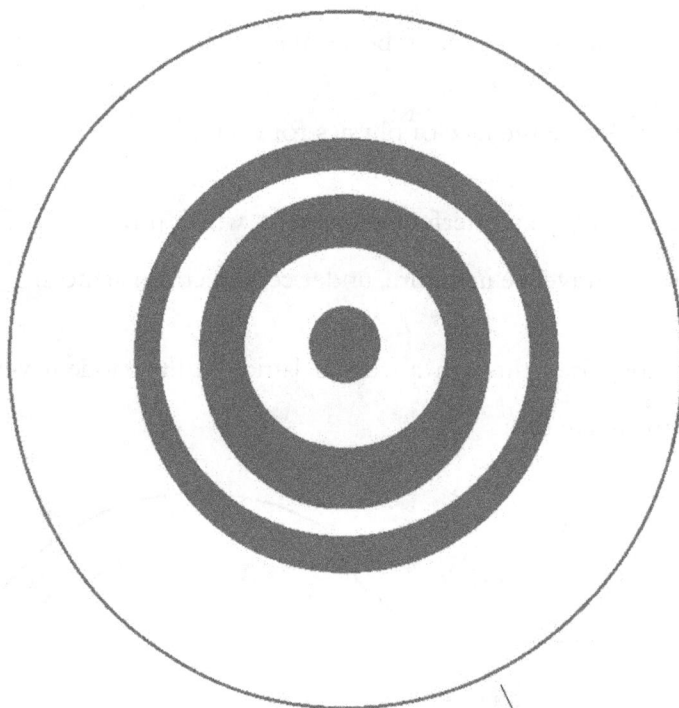

view of front of evacuated tube

The interference pattern is plain to see, albeit a circular pattern.

The Search for Classical Limitation

One of the biggest questions still facing us in the quantum world is – what is the limit in size that quantum behaviour can be shown to work?

In other words, we don't ever seem to interfere, do we?

Why not?

There has been considerable investigation into just how far we can increase the size of an object to see at what stage quantum behaviour breaks down.

We now examine this.

Now, electrons and quarks are considered to be fundamental particles, that is, they can't be broken down into smaller particles.

However, protons and neutrons are not fundamental particles.

The next step in climbing the ladder to making larger things interfere came in the 1940s, when **neutrons** were made to interfere.

This was carried out in the Oak Ridge National Laboratory in the U.S.A.

Then we had to wait until 1999 for a really large particle to be made to interfere.

This was carried out in Vienna University in Austria.

What was made to interfere here?

The answer is that a **carbon 60** molecule was made to interfere.

This consists of 60 carbon atoms arranged as **buckminsterfullerene**.

This is named after a structure which was designed by architect R. Buckminster Fuller.

Since then even bigger things have been made to interfere.

One of these is the **fluorinated fullerene molecule**, which is $C_{60} F_{48}$.

Note that here, the numbers in the formula refer to the **number of atoms in each molecule** as opposed to the atomic number.

This molecule has a mass of some 1632 times the mass of a proton, so that this molecule is quite large.

The next progression was a long chain molecule based on the **azobenzine molecule, $C_{12} H_{10} N_2$.**

This molecule, although smaller than the fluorinated fullerene molecule is larger since a very long chain molecule was produced which was based on the azobenzine molecule.

What came next?

The most amazing piece of interference to date (of writing) is the interference which was produced of a **quantum machine**.

What is a quantum machine?

This is based on a piece of aluminium which is 50 micrometres in length.

Physicists in the University of California, in Santa Barbara have built it.

It is a great step towards a so called 'classical object', in that it can be seen with the naked eye.

The aluminium was made to rest in its ground state by cooling it down to near absolute zero, and then, by bringing a tiny oscillating electrical circuit near to it, it was made to **oscillate in a superposition of two states at the same time**.

That is, it was seen to be completely at rest and to oscillate at the same time.

It can be represented like this:

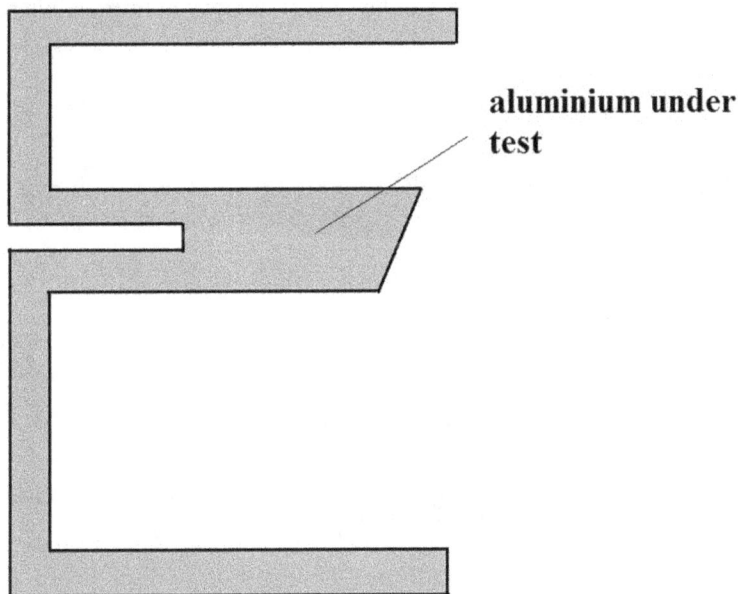

aluminium under test

Actually, in principle golf balls should be able to interfere, and even we should be able to do it.

Nobody has yet come up with a satisfactory explanation as to why this is not the case.

In fact, we can be thought of as consisting of waves.

Planck Scales

Actually, the universe does not run smoothly.

For example, time does not flow smoothly, but instead it is grainy.

There is a smallest amount of time, smaller than which it is impossible to achieve.

This is called the Planck Time

This is related to the smallest possible amount of distance which it is possible to achieve.

This is called the Planck Distance.

The Planck Distance is given as **1.616 x 10^{-35} m**.

It is impossible to quote any distance smaller than this.

The Planck Time is given as the time taken for a photon to pass through the Planck Length.

Calculate this distance now.

The Planck Time is given as **5.391 x 10^{-44} s**.

It is impossible to quote any time shorter than this.

In fact the wavelength of large objects can be found from the famous de Broglie equation, thus:

$$\lambda = \frac{h}{mv}$$

Where λ is our wavelength, **h** is the Planck constant, **m** is our mass, and **v** is our velocity.

Example

A snooker ball has a mass of **0.17 kg** and is moving along a snooker table at a speed of **0.3 ms^{-1}**.

If the value of the Planck constant is **6.6262 x 10^{-34}Js**, calculate its de Broglie wavelength.

Solution

$$\lambda = \frac{h}{mv}$$

$$\lambda = \frac{6.6262 \times 10^{-34}}{0.17 \times 0.3}$$

$$\underline{\lambda = 1.299 \times 10^{-32} \text{ m}}$$

Questions

1. Calculate the de Broglie wavelength of a **3 kg** bird which is flying at a speed of **1.0 ms^{-1}**.

2. Calculate the speed of a runner of mass **60 kg**, running at a speed of **5 ms^{-1}**.

   How do you reconcile your answer given that the Planck length is **1.616 x 10^{-35} m**?

Chapter 14: Single Photon Firing

What would we expect if we fired photons, one at a time, through a single slit?

We might expect the same result as if we had fired bullets through a single slit – all the photons having gathered in just one place.

Like this:

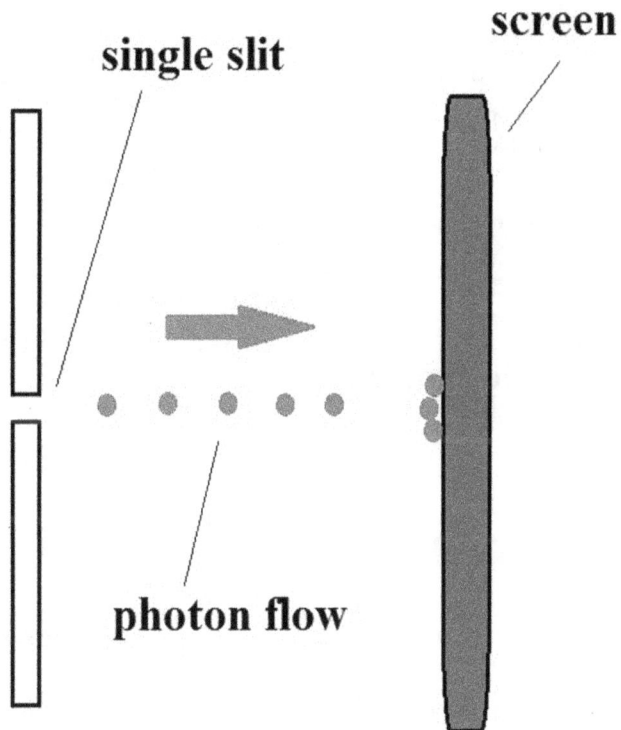

In fact this is exactly what we would get.

Now, if we fired bullets through a double slit, we would expect to get two areas where the bullets landed.

However, in the case of single photons being fired through two separate slits, we would not get two areas where the photons landed.

Now, remember that the photons are being fired one at a time, and not together.

This is quite easily achieved in practice.

Instead we would get an interference pattern set up with the double slit arrangement of photons being fired one at a time.

How is this possible, since no photon would be able to interfere with another photon on its journey?

The pattern observed would be something like this:

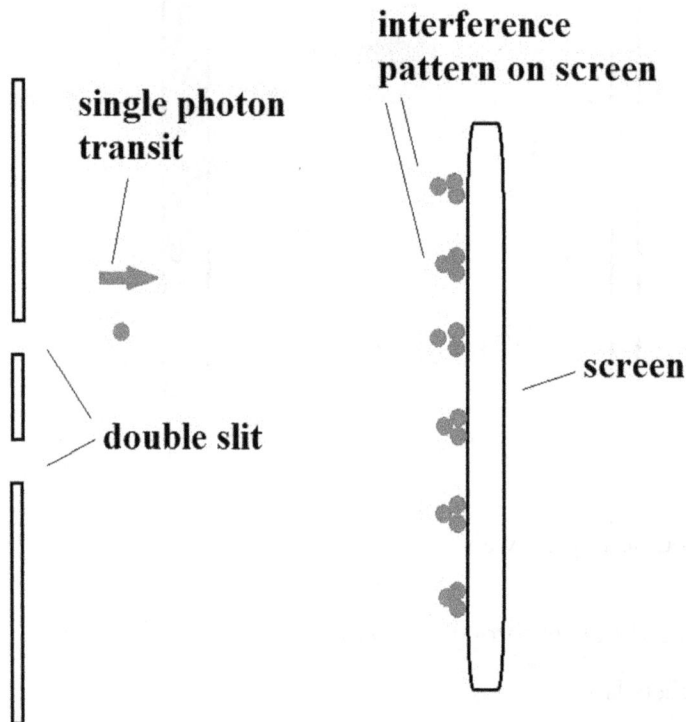

Now here is the question we need to ask ourselves:

If the photon went through just one slit, what did it interfere with?

The answer lies in the fact that, as we will see later on, unobserved quantum particles are usually in a multiple of places at the same time.

If this is the case, it could have interfered with itself during its trajectory!

Robert Austin and Lyman Page of Princeton University used a Hamamatsu photon counting camera to perform this experiment.

They achieved exactly the same results, with an interference patter being observed, even though the photons were sent individually.

Chapter 15:   The Wavelength of Light

Interference of light can also be achieved by using a diffraction grating.

What is a diffraction grating?

A diffraction grating is a piece of equipment that does not have just two 'slits' on its surface, but many.

Let us now derive a simple equation for calculating the wavelength of light.

The diagram which follows will show two slits in a diffraction grating, which in reality will be extremely close together – much closer than the scale shown in the diagram.

Thus an array of bright and dark fringes may be observed by the use of this piece of equipment.

Look carefully at the following diagram:

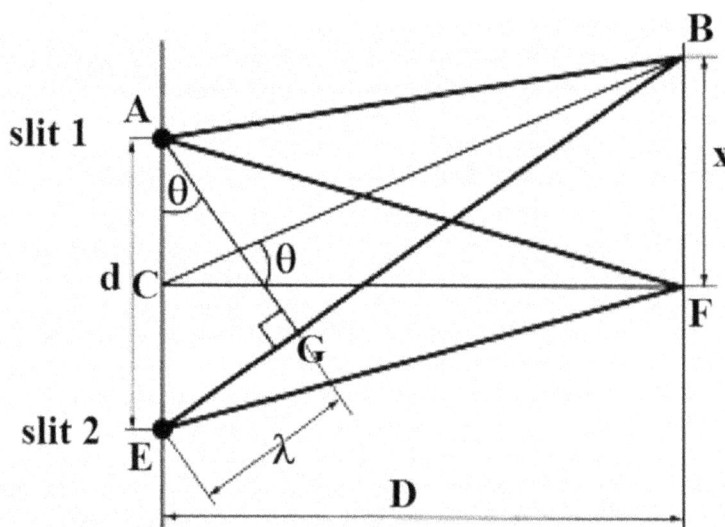

The diagram shows **slit A** and **slit B**, which are two adjacent slits in a diffraction grating, separated by a distance **'d'** metres.

'D' is the distance between the grating and the screen, and **'F'** is the distance between adjacent interference fringes on the screen.

Look now at triangles **CBF** and **AEG**.

Can you see that these triangles are similar?

This being the case, then:

$$\tan \theta = \frac{x}{D}$$

$$\tan \theta = \frac{\sin \theta}{\cos \theta}$$

$$\tan \theta = \frac{\lambda/d}{\cos \theta}$$

$$\lambda = \frac{d \, x \, \cos \theta}{D}$$

Now, since the angles involved are extremely small, then:

$$\cos \theta \text{ is approximately equal to } 1$$

Hence the equation reduces to:

$$\lambda = \frac{d\,x}{D}$$

We can now use this equation to calculate the wavelength of light as determined from using a laser beam and a diffraction grating as follows:

The experimental set up can be described by the following diagram:

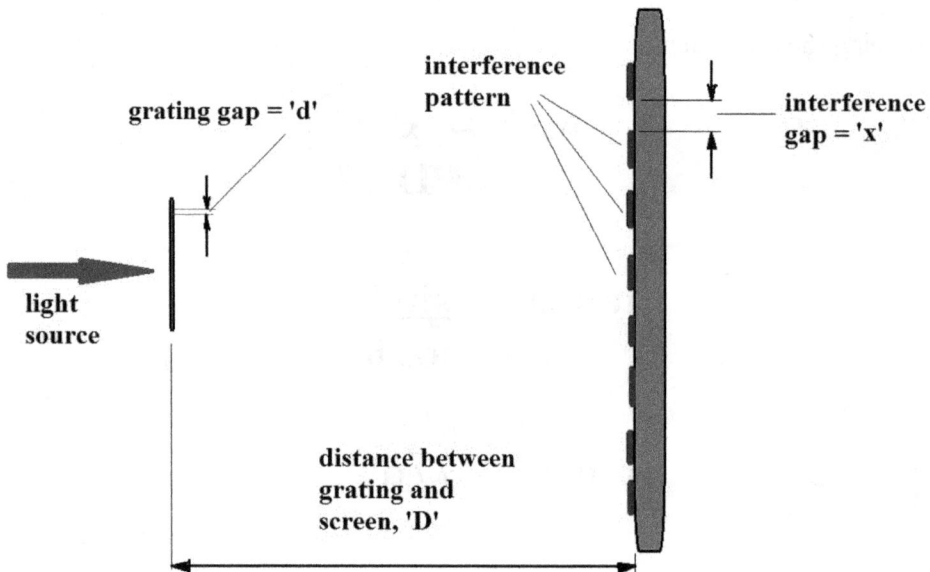

In this experimental set up the wavelength of the light passing through the diffraction grating, and as we have seen, the equation we shall use is given as follows:

$$\lambda = \frac{dx}{D}$$

Where the symbols have the meanings indicated in the diagram.

Example

A diffraction grating has 2000 lines per centimetre.

    (a) Calculate the gap between gratings lines, 'd'

    (b) If the distance between interference fringes, 'x', is 0.2 metre, and the distance between the grating and the screen, 'D', is 1.67 m, calculate the wavelength of the laser light that is involved.

    (c) What colour is this?

Questions

    1.   A diffraction grating has 4000 lines per cm.

    (a) What is the distance between each slit?

    (b) If the distance between each interference fringe, 'x', is 0.3 m, and the distance between the grating and the screen, 'D', is 1.875 m, calculate the wavelength of the laser light used.

    (c) What colour is this?

    2.   If laser light of wavelength 5.0×10^{-7} m is used to generate an interference pattern through a diffraction grating:

(a) Calculate the number of lines per cm. in the diffraction grating, given that the distance between the grating and the screen, 'D', is 4.25 m, and that the gap between interference fringes, 'x' is 0.45 m.

(b) What approximate colour is the light used?

Chapter 16:   Single Electron Experiments

This experiment was originally attempted by Marton in 1952, and then by Mollenstedt and Ducker in 1956.

In this experiment **electrons** were fired, singly and individually, at a double slit.

What sort of result would this give us?

Well, if electrons are truly quantum particles and they can **'choose'** to act as waves under certain circumstances, then we should get an interference pattern!

Initially the top slit was **covered**, as seen in the following diagram:

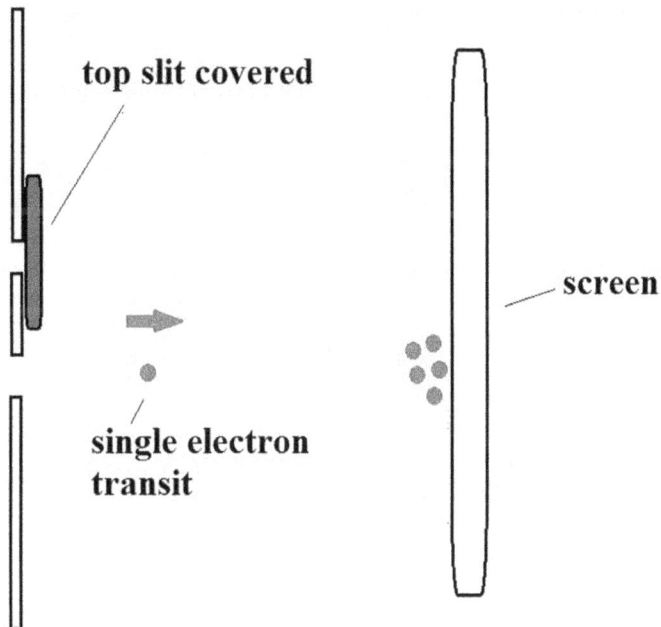

As one would expect, the electrons accumulated behind the single open slit as shown.

Next, the bottom slit was covered so that no electron could pass through it.

As one would expect, the electrons accumulated behind the open top slit, as shown below:

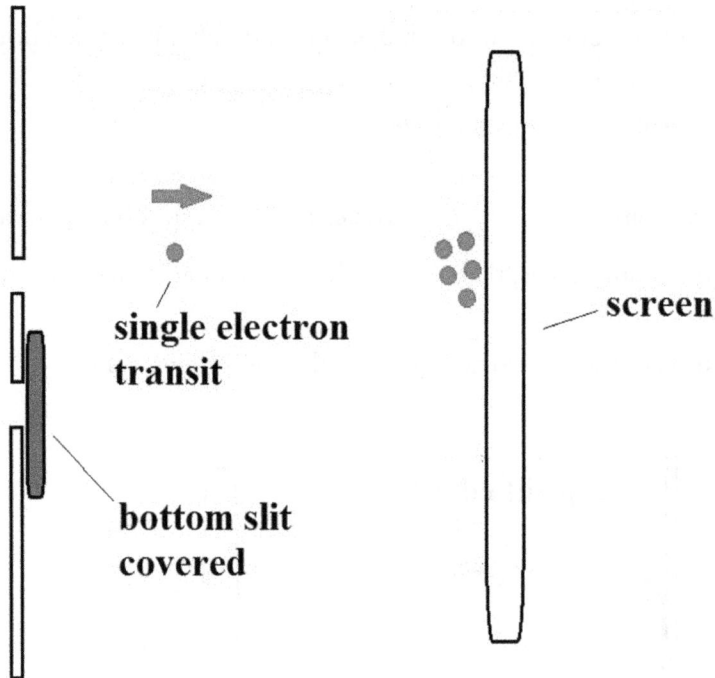

What would be the next stage in the process?

If you guessed to pass the electrons, one at a time through the equipment with both slits open, you would be correct.

This was done as shown below:

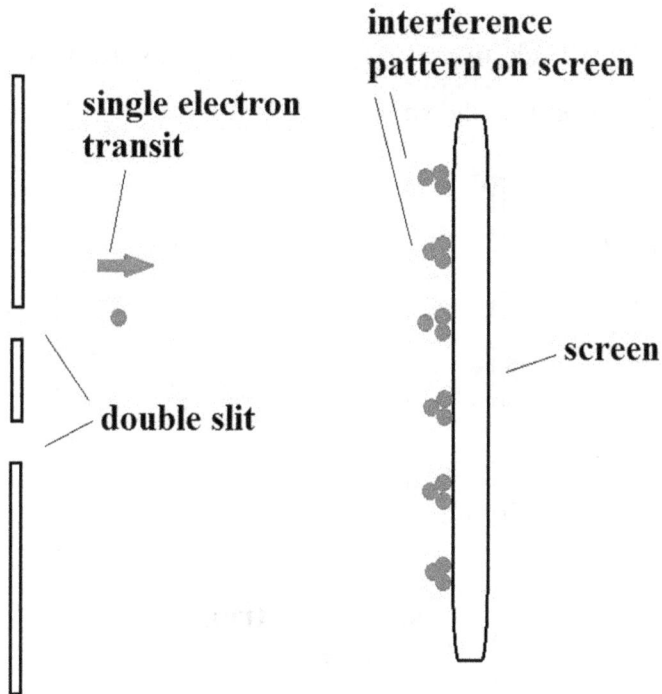

As you probably correctly guessed, an interference pattern was observed as shown above.

But wait a minute – electrons have a mass of **1.90956 x 10^{-31} kg, and they have a spin and a theoretical diameter of some 1 x 10^{-18} m**, so how could they possibly interfere?

The answer is that they also can behave as waves if they so 'choose'.

Now, remember that only one electron was fired at any one time, so what did each electron interfere with?

The answer is the same as before – unobserved quantum particles occupy multiple places at any one time, as we will see later.

Tricking the Electron

Attempts were now made to cover up one of the slits **after the electron had passed through the slit**, as shown below:

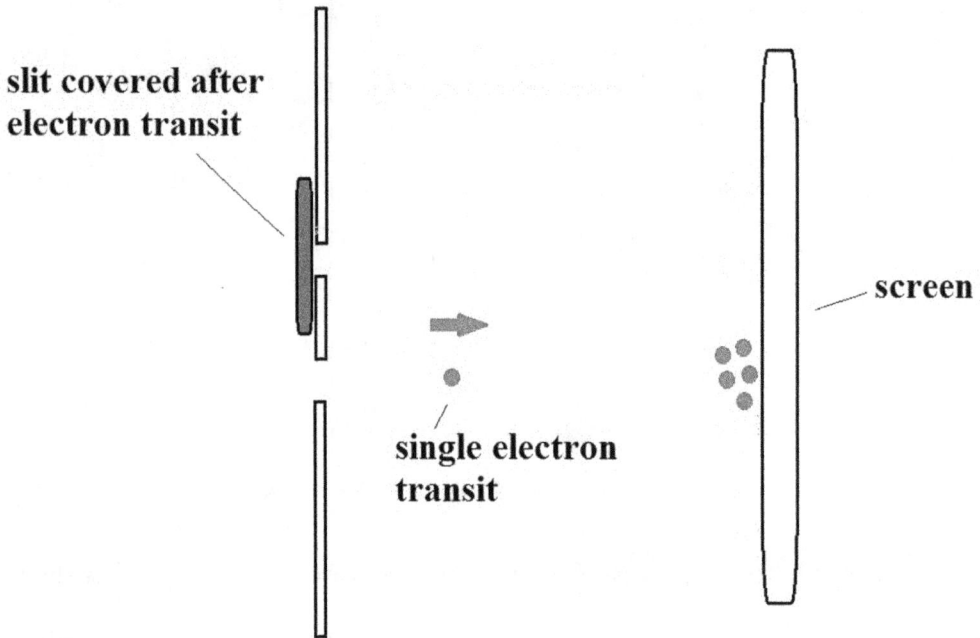

slit covered after electron transit

screen

single electron transit

It didn't matter which slit was covered up, the electron always refused to interfere if this happened.

If the electron 'thought' that attempts were being made to observe it, it would never behave in a wavelike manner, but always as a particle.

Why was this?

And further, how did it 'know' that the slit had been covered up **after** it had made its pass?

Chapter 17:   The Delayed Choice Experiment

John Wheeler (1911 – 2008) was professor of physics at Princeton University.

He was also the PhD supervisor of the eminent theoretical physicist Hugh Everett 3rd.

John Wheeler was the inventor of the terms **black hole** and **worm hole**.

In the 1980s John Wheeler devised the now famous **Delayed Choice Experiment**.

In this experiment, single photons are fired individually at a double slit as shown below:

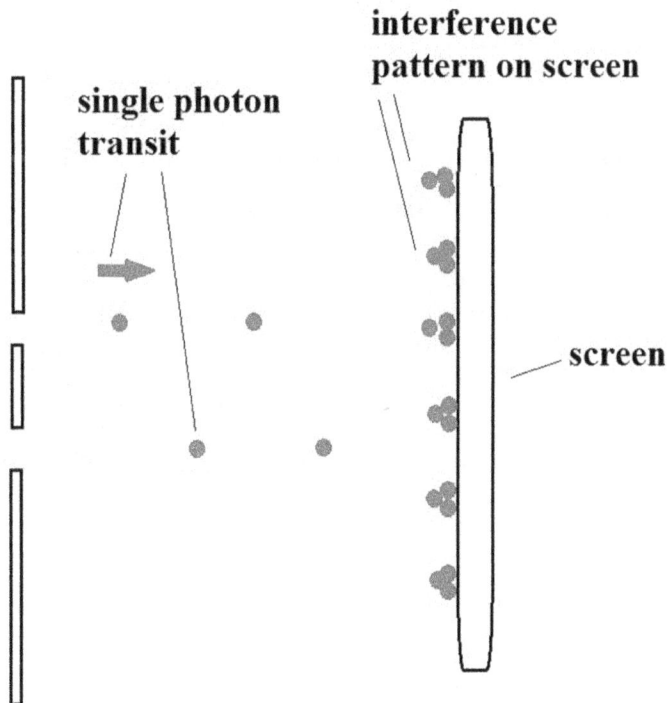

As expected, the photons formed an interference pattern as shown.

The next step in this experiment was to position detectors after the slits to try to determine which slit any photon had passed through, like this:

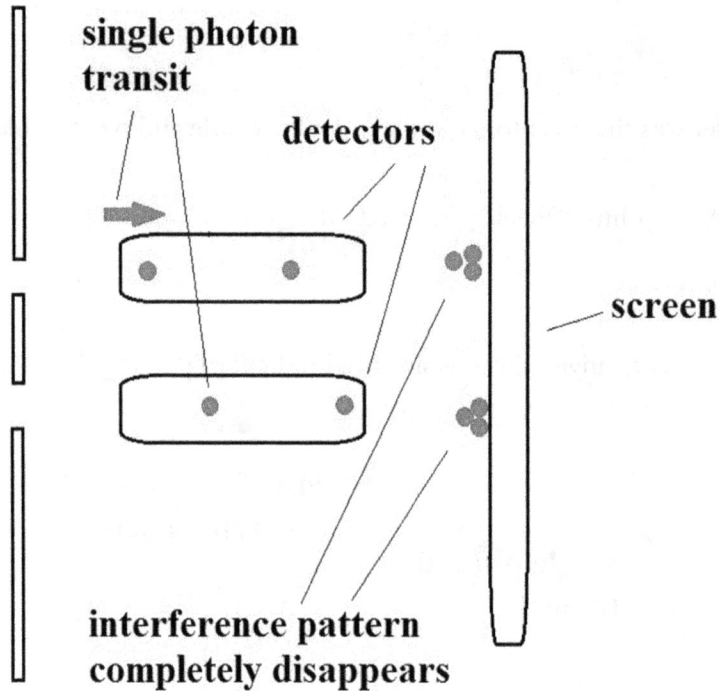

In this case, the interference pattern completely disappeared, and the photons behaved as particles, forming just two regions as shown.

This can only mean that, somehow, the photons were **'aware'** of the presence of the detectors and refused to interfere.

However, the next part of the experiment does tend to challenge belief, because here the detectors were positioned **behind** the screen, so that the photons did not pass through them, as shown below:

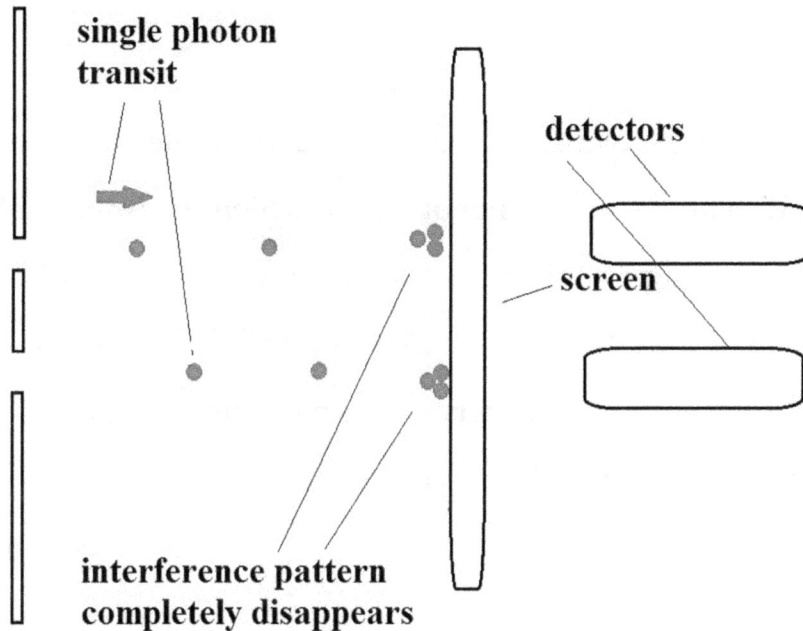

In this case, the interference pattern completely disappeared again, even though the detectors were positioned in a place **after the photons had passed through the slits**.

So how on Earth did the photons know that the detectors were there, them not being encountered until after they had passed through the slits?

The detectors were clearly in the photons **future**, so the explanation is very hard to formulate!

What is the clear implication of this experiment?

93

It must be that **actions in a quantum particle's future can affect its past!**

However, the great Richard Feynman (1918 – 1988) might possibly have shed light on this problem.

He actually said this:

> **"The basic element of quantum theory is the double-slit experiment. It is a phenomenon which is impossible, absolutely impossible to explain in any classical way and which has in it the heart of quantum mechanics. In reality it contains the only mystery ... the basic peculiarities of all quantum mechanics"**
>
> Richard Feynman

But Richard Feynman's famous **"sum over histories"** theory, for which he received the Nobel Prize in physics in 1965, can possibly shed light on the problem.

This idea suggested that quantum particles can travel, in a probabilistic sense, in all possible paths, even to the other ends of the Universe and will arrive at the destination at the same time.

This is shown below:

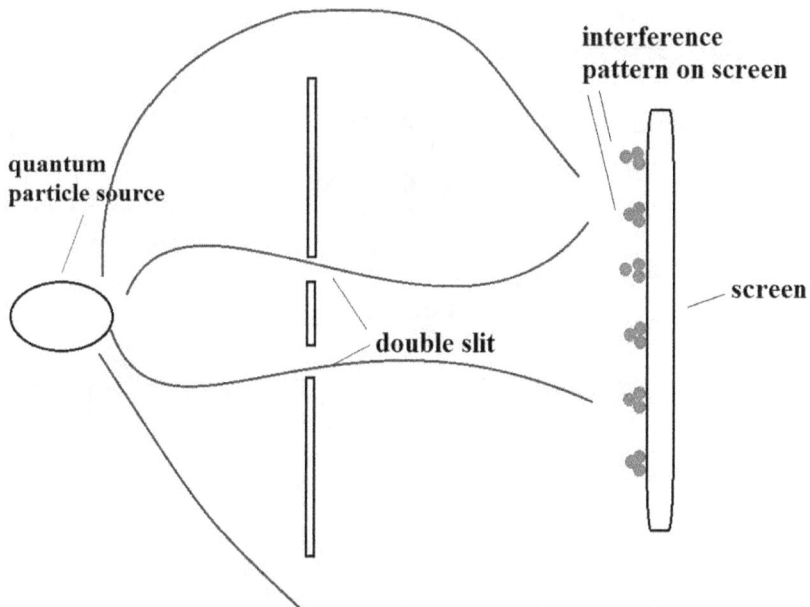

In this case, the quantum particles will travel, probabilistically, even to the other ends of the Universe and will all arrive at the destination at the same time.

This concept can go a long way to solving the problem surrounding the delayed choice experiment.

If the quantum particles travel everywhere, and it is well known that they do, and they can appear in multiple places at once, then they could somehow be aware of the detectors **behind** the screen!

Can you think of another consequence of the sum over histories concept?

It might mean that quantum particles can transit time.

Could this be true?

Think about how such a particle could traverse the Universe and arrive at the destination at the same time.

Actually, the quantum particle, probabilistically, can be a multiplicity of places at the same time, as long as it is not being observed!

The Temporal Interference Experiment

In 2005, Gerhard Paulus (et al) of the University of Texas performed an experiment which suggests that quantum particles can interfere over a time frame.

In order to do this experiment, the team devised a way to eject electrons **at different times** to see if they would interfere.

In order to achieve this, a sinusoidal laser pulse was transmitted through a cloud of **argon atoms**, as shown below:

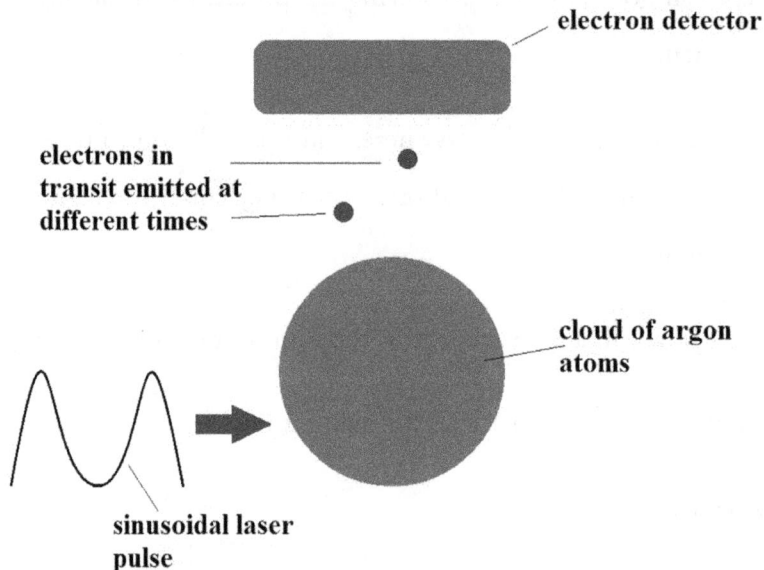

electron detector

electrons in
transit emitted at
different times

cloud of argon
atoms

sinusoidal laser
pulse

Each pulse peak, as shown above, will ionise an atomic electron.

This means that the electrons will be emitted at different times from each other.

An interference pattern was detected by the detector, thus showing that the electrons were traversing time and yet were able to interfere!

Chapter 18:   Quantum Entanglement

It is possible to create two quantum particles that are **entangled**.

This means that their reality is smeared out and their spins are indeterminate.

However, if the spin of one of the particles is measured, the other particle will adopt the opposite spin **immediately**, regardless of how far away those particles are separated.

Can you see a problem with this?

The problem that Albert Einstein, Nathan Rosen, and Boris Podolski saw was one involving the problem of superluminal information exchange, which would normally be impossible for any particle that had mass.

The Famous EPR Gedankenexperiment

Albert   Einstein,   Nathan   Rosen,   and   Boris   Podolski   produced   this Gedankenexperiment (or thought experiment), in 1935.

The purpose of it was to show that the quantum theory was an incomplete theory and that there were hidden variables that we had not yet discovered which would show the theory to be more on classical physics lines than believed at the time.

It must be stated now, that the idea of this famous experiment has been shown to be in error many times by various experimenters, the first of whom was the famous and talented French physicist Alain Aspect.

Basically, the thought experiment, (there not being the technology at the time to test it), stated that two entangled quantum particles of known mass, and both of which were moving in opposite directions as shown below, could easily have their momentum measured:

If the momentum of quantum particle 2 were to be measured, then the momentum of quantum particle 1 would be known immediately.

Further, the position of quantum particle 2 could be determined without difficulty, thus giving the exact opposite position of quantum particle 1, as shown below:

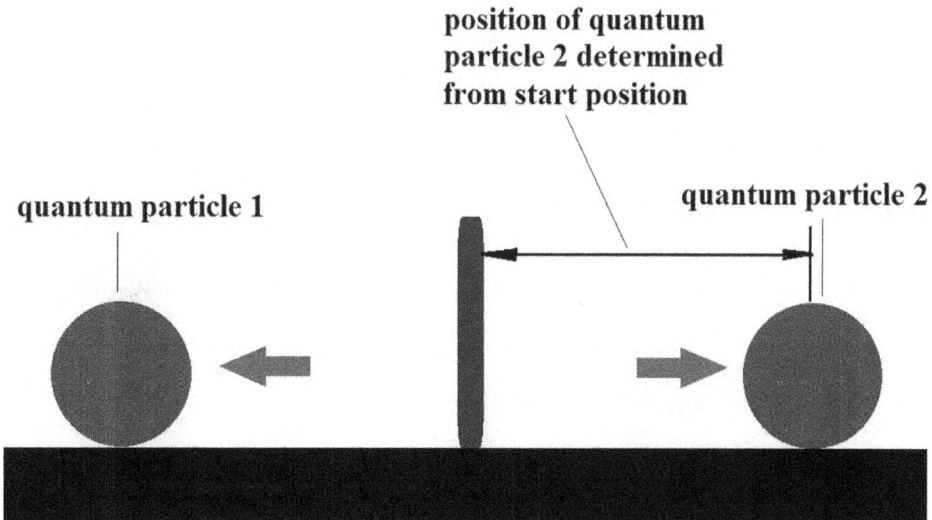

If the position of quantum particle 2 were to be known precisely, then the position of quantum particle 1 would be known immediately, by the conservation of kinetic energy (assuming no losses).

Their proposition was thus:

"Since both the momentum and position of one particle (the one on the right) can be known without disturbing it, then both quantities can be regarded as real.

But quantum theory implies that the two cannot be real at the same time.

This shows that something is wrong with the quantum theory"

But remember that their philosophy has been proved erroneous many times since.

However, there is one suggestion that could point out as to where this famous thought experiment went wrong, and that is that the classical physics equations of momentum were **incorrectly** applied to quantum particles, although it must be stated that high energy electromagnetic photons do possess momentum.

The thought experiment tried to show that the quantities that were being measured were **real** in the sense that the properties must have been attributed to the particles **before transmission**.

This is now known to be completely false when applied to entangled quantum particles, since their reality is 'smeared out' and their spins, and so forth, have not, at the transmission stage, been allocated.

This is now known to be completely true due to the experiments that we will outline later.

The three men said this:

The result of a measurement performed on one part A of a quantum system has a non local effect on the physical reality of another distant part B, in the sense that quantum mechanics can predict outcomes of some measurements carried out at B; or...

Quantum mechanics is incomplete in the sense that some element of physical reality corresponding to B cannot be accounted for by quantum mechanics (that is, some extra variable is needed to account for it)

Chapter 19:   Enter John Bell

John Stewart Bell (1928 – 1990) was a British physicist from Northern Ireland.

In 1964 he wrote a paper entitled "on the Einstein – Podolski – Rosen paradox".

He produced a mathematical inequality, which, if the Gedankenexperiment was true then the inequality would be verified.

However, if the inequality was violated, then the EPR thought experiment would be proved to be incorrect.

However, in the many experiments that have been done on the inequality, it has always been violated, showing beyond doubt that Einstein, Podolski and Rosen were wrong in that an unobserved quantum pair of particles did indeed have an indeterminate history and were totally entangled in the quantum sense.

And they were wrong about their belief that the spins and so forth of the particles were attributed to them upon creation, because these men believed, incorrectly, that the properties of entangled quantum particles were attributed **a priori**.

The proposed experimental set up for the Bell's inequality experiment is shown below:

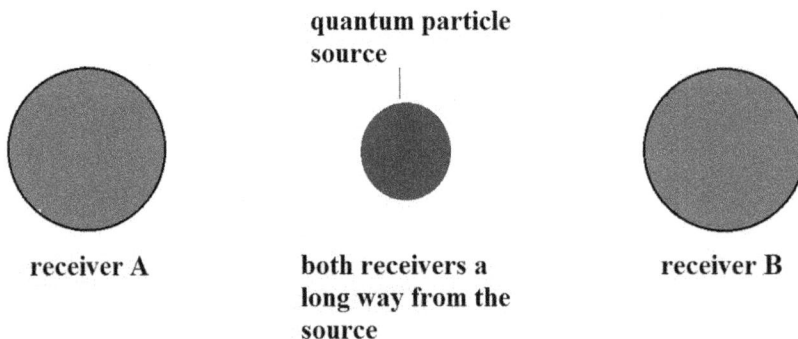

quantum particle source

receiver A   both receivers a   receiver B
                    long way from the
                    source

The source transmits entangled quantum particles (usually photons) to receivers A and B which are a long way from the source.

The distance away is important so that it can be established that the connection between the entangled pairs, once the spin has been measured on one particle, and the other entangled particle's spin is subsequently measured, has occurred before light can have been transmitted between the particles, proving that the entangled nature of the quantum particles in instantaneous and does not depend upon the transmission of information at light speed.

There are many ways to formulate Bell's inequality, but we will now consider just one of these ways.

The experimental set up always requires two observers A and B (sometimes referred to as Alice and Bob).

The detectors are always situated a long way from the source of entangled quantum particles (so that superluminal transmission can be ruled out).

Both receivers can measure whether or not the spin of the quantum particle they are dealing with is either **spin up** or **spin down**.

Both A and B can choose whether to measure spin up or spin down, and remember that many millions of entangled photons are dealt with so that a good correlation can be built up over a period of time.

For the benefit of the experiment it is assumed, as Einstein et al believed, that the spin has been attributed to the particles at the point of transmission.

If this were true, then Bell's inequality would hold.

If Bell's inequality is violated, then Einstein et al would be proved to be wrong.

Detector A can choose to measure, say, spin up (we will call this A_1, or it can choose to measure spin down (we will call this A_2).

Similarly detector B can choose to measure either spin up (we will call this B_1), or spin down (we will call this B_2).

So that A_1, A_2, B_1 and B_2 can have values of either **+1** or **-1**.

Let us now define a variable, **S**, such that:

$$S = A_1(B_1 + B_2) + A_2(B_1 - B_2)$$

It can be seen that this variable contains all possible permutations, such that either $(B_1 + B_2)$ must be zero, or $(B_1 - B_2)$ must be zero.

This can be easily checked by inserting **+1** and **-1** for all possibilities in the above part expressions.

And then the other term, which isn't equal to zero, must be equal to either **+2** or **-2**.

This again can be easily checked by inserting all permutations of **+1** and **-1** in the variable.

So that **S** must have maximum and minimum values of **+2** and **-2** respectively.

Hence the **average** value of **S** must lie in between **+2** and **-2** as shown below:

$$-2 \leq (S) \leq +2$$

This is Bell's inequality and if the quantum theory is wrong, and the spins really are allocated at photon production, then **S** will always yield values of between **-2** and **+2**.

However, if the quantum theory is true, then **S** will yield values regularly of **above +2** and **below -2**.

The main point here is that if the quantum theory is true, then the inequality will be regularly violated and the photon spins will not have been allocated at quantum particle creation, but the particles will propagate through space with **all possible histories**.

This means that their spins would be indeterminate until a **measurement** had been made on one of the particles.

Now, Bell's inequality is as follows:

$$S = A_1(B_1 + B_2) + A_2(B_1 - B_2)$$

Expanding this, we have:

$$S = A_1B_1 + A_1B_2 + A_2B_1 - A_2B_2$$

Let us now examine, for example, the case where **S** yields values of **+4** and **-4**.

What would have to happen to give us a value of **S** of **+4**?

In this case, the following would have to have the value of **+1** ......

$$\mathbf{A}_1 \ \mathbf{B}_1$$

And
$$\mathbf{A}_1 \ \mathbf{B}_2$$

And
$$\mathbf{A}_2 \ \mathbf{B}_1$$

And
$$\mathbf{A}_2 \ \mathbf{B}_2 \ \text{would have to have the value of } \mathbf{-1}$$

What problem occurs with this?

The problem is that $\mathbf{A}_1$, $\mathbf{A}_2$, $\mathbf{B}_1$, and $\mathbf{B}_2$ would all have the value of **+1** in the first three terms.

But $\mathbf{B}_2$ would need to assume the opposite value of **-1** when it was paired with $\mathbf{A}_2$ in the last term, and this could only occur if there were **instantaneous communication between the photons upon measurement**!

So that the photon giving the result of $\mathbf{B}_2$ **is equal to -1** would need to know instantaneously that $\mathbf{B}_2$ had assumed the value of **+1** when paired with $\mathbf{A}_1$.

This means that detector **B** would be somehow able to see **instantaneously** what detector **A** did and make the correct corresponding choice.

Of course, detector **B** does not see what detector **A** does, but the consequential proof is the quantum theory is correct.

This can be checked by working out a similar sequence for the result of **S is equal to -4**.

In many experiments involving entangled quantum particles (usually photons), violation of Bell's inequality has **always** been observed, thus proving that Einstein et al were completely wrong with the EPR challenge.

The particles always assume the correctly related spins when the spin on one of the particles has been observed, regardless of how far away the particles are from each other, showing that there appears to be a superluminal communication between them.

However, the process is not superluminal, but is a quantum theoretical consequence of entanglement, and as yet, we don't understand how it happens!

Einstein called this phenomenon "spooky action at a distance", and he wouldn't accept the truth of it.

The Latest Violation of Bell's Inequality (at time of writing)

This was conducted by Salart et al in Switzerland.

What did this involve?

Two entangled photons were sent from Geneva to the towns of Satigny and Jussy which are some eighteen kilometres apart.

The actual distance travelled by the photons was some 17.5 kilometres in each case.

This is shown below:

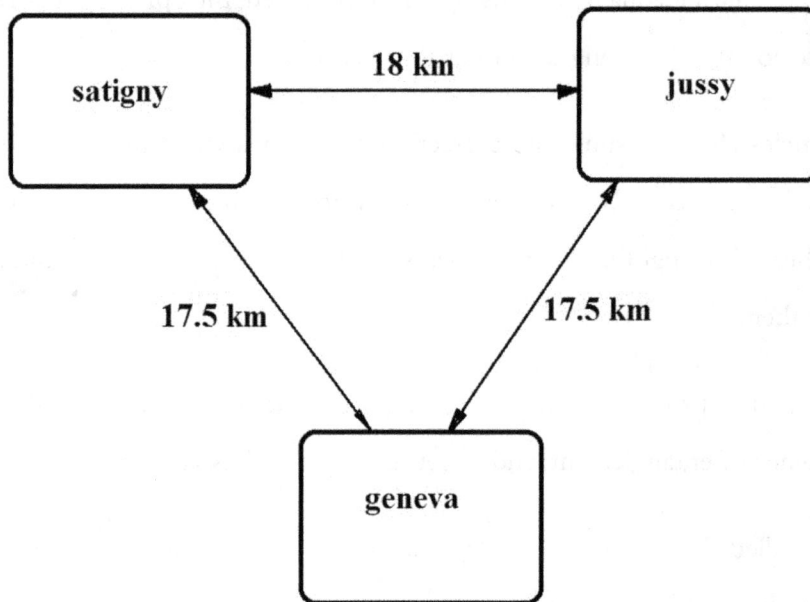

What was the result?

Bell's inequality was consistently violated, with confirmation that the assumption by the other photon of the correct opposite spin took place before light could have made any communication.

Chapter 20:   Retrocausality

What is retrocausality?

It is the phenomenon that would occur if an event in a quantum particle's future could affect that quantum particle's past.

One very good way to investigate this is by an experiment proposed by John Cramer of Washington University, Seattle.

Here is the outline of what has been proposed:

Retrocausality Experiment

John Cramer of Washington University, Seattle, has proposed a very interesting experiment to the American Association for the Advancement of Science.

In this proposed experiment, two entangled photons are separated and each is subsequently made to interfere in its own double slit.

If observation is made by looking at the exit of the double slits, the photons will detect being observed and will not interfere.

In this case, it will behave as a particle.

However, if the observation is made away from the double slits, the photons will behave as a wave and will interfere.

However, there is a time delay in this experiment.

The photon which is observed is in the **future** compared with the unobserved particle.

The twin photon should adopt the same identity as the observed particle – to either behave as a particle or as a wave.

However, the big question is "can the non-observed photon receive the entangled message in the past compared with the observed photon?

How is the observation made in the future compared with the twin photon?

The answer is that the observed photon must pass down some 10 km of fibre optic cable before being observed, whereas the twin photon does not have a time delay.

The experimental set up is as follows:

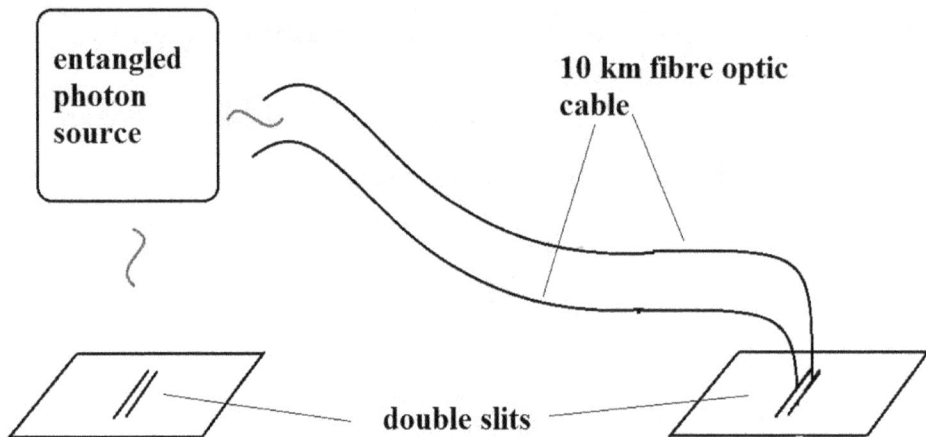

The results of this experiment will be of great interest to us.

Chapter 21:   The Heisenberg Uncertainty Principle

The Heisenberg Uncertainty Principle is given below:

$$\Delta p \Delta x \geq \tfrac{1}{2}\hbar$$

Where Δp is the uncertainty in the momentum of the quantum particle and Δx is the uncertainty in the displacement of the quantum particle.

And $\hbar$ is the reduced Planck constant, and is given as **1.054593 x 10^{-34}Js**

And remember, that since $\Delta p = m\ \Delta v$, then the speed of the quantum particle is uncertain too.

However, there is a traditional error that needs to be corrected about the definition of the Heisenberg Uncertainty Principle.

What is this traditional error?

It is that the equipment that is being used causes the uncertainty.

This can be represented thus:

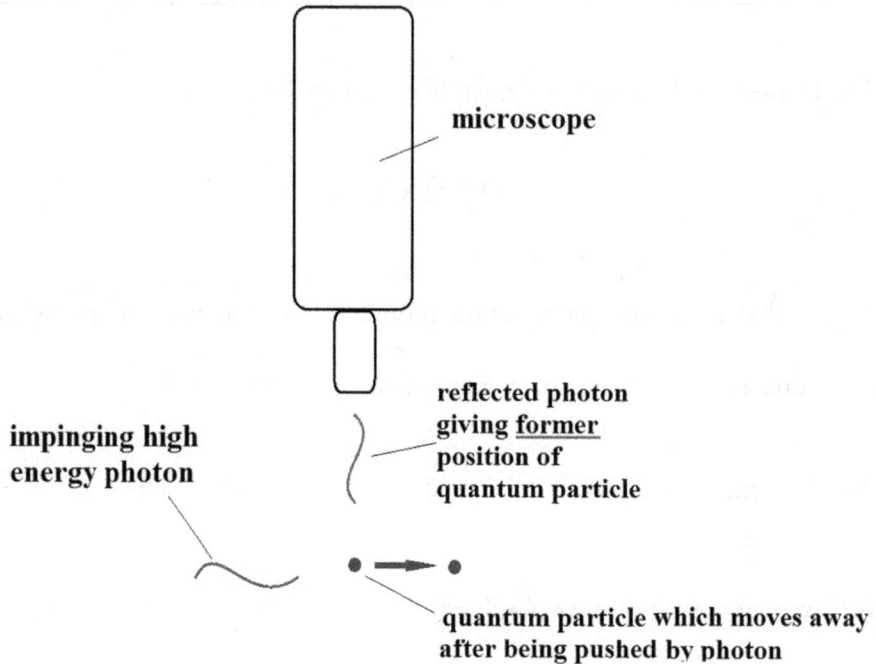

microscope

reflected photon giving <u>former</u> position of quantum particle

impinging high energy photon

quantum particle which moves away after being pushed by photon

In this view, it is the result of the impinging high energy photon upon the quantum particle that causes the uncertainty in both the position and speed of the quantum particle.

However, this is totally incorrect.

What the uncertainty principle means is that, without any measuring equipment now, impinging high energy photons, the speed and position of the quantum particle is uncertain, per se, without any external interference at all.

In other words, we do not know, (except that we can have a probabilistic indication) as to where a quantum particle is or what its speed is.

However, if we know one of the quantities with some certainty, the consequence for the other of the two uncertain quantities is that it has an even greater uncertainty.

In other words, an unobserved quantum particle can appear in a multiplicity of places at once, and can have a variety of speeds at any one time.

This is certainly counterintuitive, but is absolutely true of the quantum world, the workings of which we do not really understand!

Let us now consider an example, by way of illustration:

Example

An unobserved electron is considered to be nearly at rest with an uncertainty in speed of a billionth of a metre per second.

What is the uncertainly in its position?

Notice that this quantum particle does not have any impinging high energy photons approaching it, nor is it in the vicinity of any measuring equipment!

Solution

$$\Delta p \Delta x \geq \tfrac{1}{2}\hbar$$

$$m \Delta v \Delta x \geq \tfrac{1}{2}\hbar$$

Taking the electronic mass to be **9.109565 x 10^{-31} kg**, we have:

$$(9.109565 \times 10^{-31}) \times (1 \times 10^{-9}) \, \Delta x \geq \tfrac{1}{2} (1.054593 \times 10^{-34})$$

Hence, the uncertainty in position is given as:

<u>57,884 m or 58 km</u>

In other words we could say that the quantum particle wasn't there at all!

And had the uncertainty in speed been any smaller, the uncertainty in position would have been even greater!

We could say that the uncertainty principle isn't anything to do with uncertainties in measurement, but to do with uncertainties in **existence** (from the point of view that we wouldn't know where the quantum particle was!).

It is certainly true that the phenomenon of **quantum tunnelling** can be taken to mean that the quantum particle is somehow at the other side of the classically impenetrable barrier, as shown below:

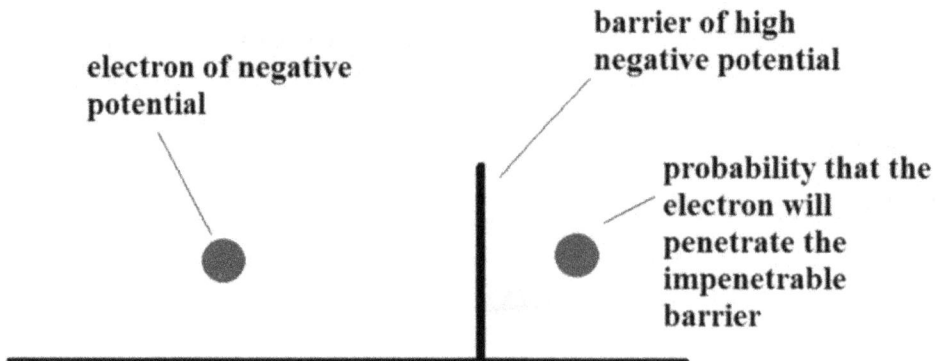

electron of negative potential

barrier of high negative potential

probability that the electron will penetrate the impenetrable barrier

Actually, no alpha particle can possibly possess enough energy to overcome the action of the binding energy keeping it in a nucleus, so it is barrier penetration at work there, as well as in the quantum particles in semiconductors penetrating the barriers that exist at the boundaries of the p n junction.

So we have learned that unobserved quantum particles can be in a multiplicity of places at once – and those places can, statistically, be separated by great distances!

The diagram below shows an unobserved electron (not to scale!) existing in a multiplicity of states in spacetime:

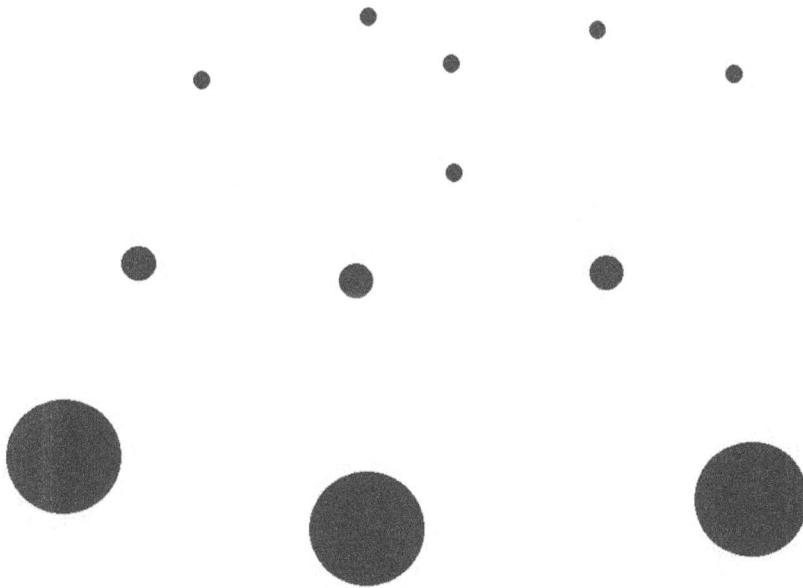

The Schrödinger equation was deduced in 1925 in order to predict the probable state in space and time of a quantum particle, and an unobserved quantum particle is clearly **'smeared out'** in spacetime.

Questions

1. An unobserved neutron, of mass $(1.674920 \times 10^{-27})$ kg has an uncertainty in its speed of one millionth of a metre per second.

   What is the uncertainty in its position?

2. An electron of mass $(9.109565 \times 10^{-31})$ kg has an uncertainty in its speed of (2×10^{-6}) ms^{-1}.

   What is the uncertainty in its position?

Chapter 22:   The Wave Function, Ψ

The wave function itself is analogous to the **amplitude** in the general wave equation.

The wave function, **Ψ**, has a very special property.

What is this property?

This property is that its **square** will give us the **probability** as to where and how many of a quantum particle is in spacetime.

The implication as to whether or not a quantum particle existed in a given place at a given time deeply disturbed Schrödinger.

As a consequence, he devised his idea of **the cat** in a box, with a phial of poison which could only be released if a single radioactive atom emitted a particle, or not.

Remember also that the emission of a radioactive particle is a **random** process, so that it could be emitted in a second or a thousand years.

He said that the cat, until observed, would be in a state of either being dead, or alive, or in some indeterminate state in between the two, or, possibly did not exist!

But it is important to remember that Schrödinger devised the idea **in order to ridicule the quantum theory** as he currently understood it at the time.

However, the analogy is now used as the standard way of describing the indeterminate state of a quantum particle (not considering that a cat would be a classical object!).

In the diagram below, the cat is represented in an indeterminate state (not alive nor dead), the radioactive source is able to release just one alpha particle which will hit the detector and cause the heavy weight to break the phial with poison in it.

However, we have no idea when the alpha particle will be emitted:

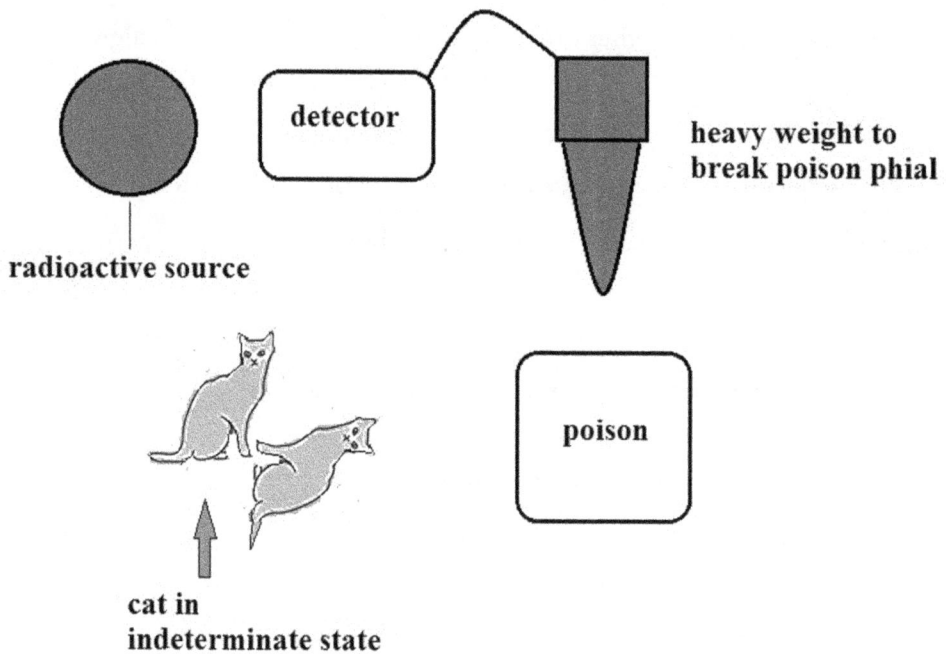

detector

heavy weight to
break poison phial

radioactive source

poison

cat in
indeterminate state

Chapter 23:   The Copenhagen Interpretation

In the Copenhagen meetings in the 1920s, many of the leading people in the quantum theory met, and out of those meetings came what is known as the 'Copenhagen Interpretation'.

What were the main tenets of the Copenhagen Interpretation?

They included the following:

That one should accept that the equations of quantum theory work perfectly but that one should never ask how they work

That a measurement on a quantum system would **collapse the wave function, ψ**

What are the implications of the last point?

They are that an observer, whether a conscious being, or a detector, is **not a quantum system, but is a classical object**

This means that there is a big difference between quantum and classical systems

At what size do quantum systems become classical objects?

At the moment, this is unknown

What is the main difference between a quantum state and a classical state?

The difference is that objects in a quantum state **exist in a superposition of many states**

Of course, the best example to date is the quantum machine which we have already discusses and which can be seen with the naked eye and which has been made to assume two quantum states at the same time, these being to both oscillate and to remain perfectly at rest at the same time!

What does this imply?

This implies that the quantum machine is a **quantum system**

But to date, any bigger object has only been observed to exist in **just one state**

What does this mean?

This means that it is not in a quantum state, but is defined as being in a **classical state**

Here is another version of the Schrödinger equation:

$$\overset{\textbf{wave function}}{\frac{d^2\,\psi}{dx^2} + \frac{8\,\pi^2 m}{h^2}\,(E-V)\,\psi = 0}$$

position   **energy**   **potential energy**

Now whereas the square of the wave function, Ψ^2, tells us something about the probability as to where in spacetime the quantum particle is, remember that the whole equation describes the total distribution of the quantum object in spacetime.

Remember also that any quantum system or object **remains in a superposition of an infinite number of states at all times**

What does this imply about an unobserved quantum particle?

It implies that the quantum particle exists **in all parts of the universe and in all times** at any one instant.

The General Thrust of the Copenhagen Interpretation

The general thrust of the Copenhagen Interpretation is to say that we should only concentrate on the highly successful equations of the quantum theory and not to question the how nor the why as to exactly what is going on in the quantum world.

However, in more recent years many quantum physicists now want to try to understand what might be going on in the quantum world.

One of the main proposals that came out of the Copenhagen interpretation was that if a quantum particle is observed, then the wave function, Ψ, will collapse immediately.

However, Dr Euan Squires, formerly professor of applied maths at Durham University in the U.K., stated that the wave function could not collapse.

To quote him, he said:

> **" …. Quantum mechanics should be able to explain how the wave function reduces. In fact, however, it says very clearly that the wave function cannot reduce!"**

He, like many others at present, thought that the wave function cannot collapse.

His theory is based on a substantial mathematical basis and is not just some idea dreamed up.

Actually, the mathematical basis is quite correct.

Question

State in your own words what your opinion is of the Copenhagen Interpretation of the quantum theory.

Chapter 24:   Hugh Everett 3rd (1930 – 1982)

Hugh Everett 3rd produced his ideas in the 1950s.

His Ph.D. supervisor was the famous physicist John Wheeler, who was responsible for creating the phrases, 'black hole' and 'worm hole'.

The problem was that John Wheeler was very well acquainted with Neils Bohr, who would not accept that the wave function could not collapse.

The view of Hugh Everett 3rd was that the wave function certainly did not collapse, but was fixed, upon a quantum particle being observed in this world, but only in this world.

His original Ph.D. thesis went on to describe what is known as the **Relative State Theory**, which is also known as the **Many Worlds Theory**.

What was proposed was that the wave function was only pinned down in this world (or this Universe), and the other infinite possibilities for the existence of the probabilistic values of the wave function continued in other worlds.

This idea must not be confused with a multiverse theory.

But Hugh Everett 3rd could not accept the implication that the wave function would collapse if an observation were to be made.

He believed, as now do many others, that the Schrödinger equation, and in particular the wave function, could not be collapsed, but that all objects **still exist in the quantum state of superposition** even after the observation.

This is the main thrust of the Relative State Theory.

Can you think of a very important consequence of this?

The consequence is that the **observer** is **not** a separate entity from the quantum system

But instead the observer is **a part of the quantum system**

This means that the **whole universe is one entangled quantum system**

So what is the obvious implication of this as applied to when an observation of a quantum particle is made?

It is that we know that when an observation takes place, the location of a quantum object or system is **determined.**

However, if the Schrödinger equation is **still intact** and the wave function, ψ, **has not collapsed**, then the observer and the quantum system must have split, and the rest of the quantum system and all its infinite positions in spacetime still exist in many different worlds, although there is no explanation as to how this can be yet.

Further, the implication is that the Copenhagen Interpretation falls short of truth, and in some ways is incorrect.

However, John Wheeler, with agreement from Hugh Everett 3rd, proposed that all the references to the Relative State Theory be omitted from the thesis.

Hugh Everett agreed to this, and duly omitted all reference to it.

However, as soon as the Ph.D. was awarded, Hugh Everett published the whole thesis and put it in circulation.

What was implied was that the wave function still existed in its many forms, but in other worlds, being lost for eternity to all of us in this world.

If a decision is unavoidable (as with a measurement) the world splits and a brand new one is formed, identical to this one but with the alternative option for Ψ existing there

This means that Schrödinger's cat is **both** dead and alive with each option existing in a different world.

In fact, mathematically speaking, if the Schrödinger equation is correct, which it most definitely is, then the relative state theory is the only possible explanation for the outcome of the equation, mathematically speaking.

The many worlds theory can be expressed something like this:

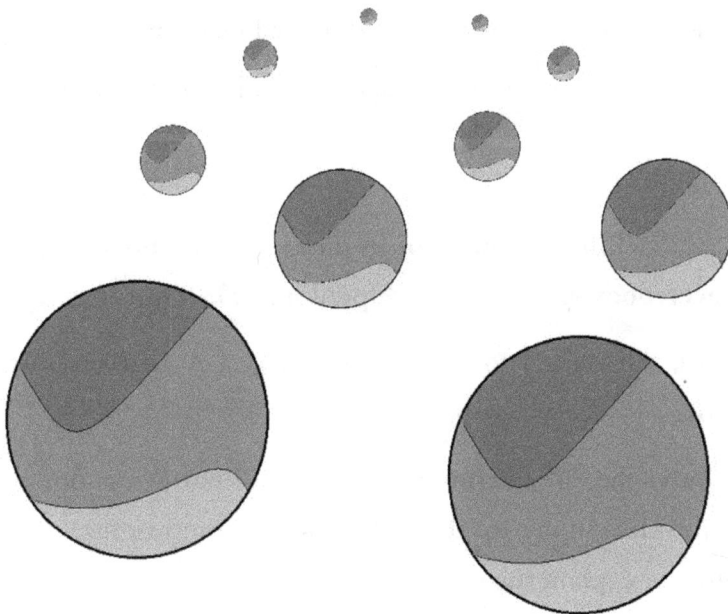

However, there remains no practical explanation as to how these many worlds can form.

Everett's theory says that the measurement determines which world we live in, and it forever separates the two worlds.

Everett's work is based on a precise mathematical foundation as opposed to a fictional guess

Many eminent physicists hold this theory as the **only** feasible explanation.

Who is David Deutsch?

David Deutsch is a world famous quantum physicist from the University of Oxford.

In 2007, David and some of his colleagues proved that some of the key equations in the quantum mechanics arise from the mathematics of the relative state theory.

David Deutsch said this:

> **"Everett was before his time, not in the sense that his theory was not timely – everybody should have adopted it in 1957, but they did not.**
>
> **Above all, the refusal to accept Everett is a retreat from scientific explanation.**
>
> **Throughout the 20th century a great deal of harm was done in both physics and philosophy by the abdication of the original purpose of those fields: to explain the world.**

We got irretrievably bogged down in formalism, and things were regarded as progress which are not explanatory, and the vacuum was filled by mysticism and religion and every kind of rubbish.

Everett is important because he stood out against it, albeit unsuccessfully; but theories do not die and his theory will become the prevailing theory. With modifications" (2)

Another quantum physicist from the University of California has said that:

"This work will go down as one of the most important developments in the history of science".

Famous Scientists Latter Views on the Quantum Theory

What did many of the famous founding fathers say about the quantum theory in later times?

Albert Einstein said:

"Reality is merely an illusion, albeit a very persistent one"

"The more success the quantum physics has, the sillier it looks. ... I think that a 'particle' must have a separate reality independent of the measurements. That is an electron has spin, location and so forth even when it is not being measured. I like to think that the moon is there even if I am not looking at it. ... God does not play dice with the cosmos"

Erwin Schrödinger said this:

"I don't like it and I'm sorry I ever had anything to do with it"

Question

State in your own words what your opinion is of the Multiple State Theory.

Chapter 25:   Heisenberg's Temporal Energy form of the Uncertainty Principle

Heisenberg's Uncertainty Principle states this:

$$\Delta p \Delta x \geq \tfrac{1}{2}\, \hbar$$

But remember that the according to Newton's second law, the rate of change of momentum equates to net force, so the principle could be quoted like this:

$$\frac{\Delta p \Delta x\; t}{t} \geq \tfrac{1}{2}\, \hbar$$

Hence this can be stated thus:

$$\Delta F \Delta x\; t \geq \tfrac{1}{2}\, \hbar$$

Remember also that the product of net force and time is equated to work done, or energy, so the principle can be eventually quoted like this:

$$\Delta E\; \Delta t \geq \tfrac{1}{2}\, \hbar$$

This version has extreme implications for the Universe.

This tells us that there can be a very small uncertainty in energy for a very uncertain time, or, for a length of time with a very small uncertainty, a huge amount of energy with a huge uncertainty.

What does this mean in terms of spacetime?

It means that, using Planck scales, we can have huge amounts of energy (as long as the time does not exceed the Planck time).

Such a time is called **"virtual time"**.

Time has to exceed the Planck time to become a real time in this Universe.

Remember that the Planck length can be calculated from the following equation:

$$\lambda_p = \sqrt{\frac{G \hbar}{c^3}}$$

Calculate the Planck length, the smallest length possible to have in this Universe, given that the values of the constants are as follows:

$$G = (6.673 \times 10^{-11}) \text{ N m}^2 \text{kg}^{-2}$$

$$\hbar = (1.054593 \times 10^{-34}) \text{ J s}$$

$$c = (2.997925 \times 10^{8}) \text{ m s}^{-1}$$

What did you get?

The Planck length is given as (**1.616 x 10^{-35}) m**.

Also, the smallest possible time that can exist in this Universe is the **Planck Time**.

This is given as the time taken for a photon travelling in vacuo to cross the Planck Length.

Work it out now.

What did you get?

The Planck Time is given as (**5.391 x 10^{-44}**) s

So as long as the Planck time is not exceeded, virtual time exists.

In virtual time, spacetime is seething with **quantum fluctuations**.

That is to say that huge amounts of energy can be drawn in virtual time to create quantum fluctuations (in the form of particle antiparticle pairs) and in the form of Lorentzian wormholes, **as long as the fluctuations disappear before the Planck time is crossed**.

This means that conservation of energy is never violated because the time in which energy is drawn from the vacuum never becomes real.

This means that spacetime is not empty, but is continually seething with quantum fluctuations all the time.

It is quantum fluctuations that cause the creation of the particle antiparticle pairs in Hawking radiation, albeit that the energy there is derived from the gravity of the black hole in question.

Questions

1. How much energy is available in a quantum fluctuation for a time equal to the Planck time?

2. By how much does this energy exceed the energy required to produce a neutron / antineutron pair for less than the Planck time?

Chapter 26:   The de Broglie Matter wave Equation

We have briefly met this before, but we will now apply this to intermediate masses.

The great Louis de Broglie suggested in 1924 that if photons could 'decide' whether to behave as waves or particles (if they 'thought' they were being observed), then maybe matter particles could also 'decide' whether or not to behave as either waves or particles.

In the Solvay conferences in the early 1920s, de Broglie was spoken to in rather harsh terms by some of the father figures of the quantum theory, which, apparently, upset him.

But as the great David Deutsch says, theories will not die, and eventually experiments were carried out that completely supported de Broglie's view that matter particles were also wave like in nature, should they not suspect that they were being observed.

The de Broglie wave equation can be quoted as follows:

$$\lambda \;=\; \frac{h}{mv}$$

Where, λ is the wavelength of the body under scrutiny, h is the Planck constant of value ($h = (6.62621 \times 10^{-34})$ Js, and m is the mass of the body, and v is its speed.

Example

An electron of mass (9.11×10^{-31}) kg has a speed of 0.3 ms^{-1}.

Calculate its de Broglie wavelength.

Solution

$$\lambda = \frac{h}{mv}$$

Hence,

$$\lambda = \frac{(6.62621 \times 10^{-34})}{(9.11 \times 10^{-31}) \times 0.3}$$

$$\lambda = (2.4245 \times 10^{-3}) \text{ m}$$

Question

If a **2 mg** snooker ball travels at a speed of **1.0 ms^{-1}**, calculate the de Broglie wavelength of the ball

Chapter 27:   Non Locality

What is non locality?

It is something that Albert Einstein and others were loath to accept up to their deaths.

Einstein called it "spooky action at a distance".

Can atoms in a distant star or galaxy have any effect on atoms here on Earth?

Yes, their gravity can have an influence on us.

The obvious example is how we are bound by the Sun's gravity field.

Can you think of something strange about the effect of the Sun's gravity on Earth?

Yes, because if we could make the Sun disappear from the Universe at the flick of a finger, the Earth would still orbit the former position of the Sun for some eight minutes.

Why is this?

It is because the mediating bosons which mediate gravity travel at light speed, and they would take some eight minutes to reach Earth.

These bosons, although to date never having been detected, are called **gravitons**.

However, we could almost forgive Einstein et al for thinking that entanglement was not correct.

The reason is that entangled quantum particles will appear to transmit data about their properties **instantaneously**.

So that a pair of entangled quantum particles, say for example, an electron and a positron, which become separated by a huge distance, will automatically seem to know their respective data, (for example concerning the spin of the twin particle), as soon as the spin on the opposite twin particle is measured.

So that if the spin on one of the quantum particles is measured (to be say spin up), then the twin particle will adopt spin down immediately notwithstanding the distance of separation.

Is this counterintuitive?

According to the way our brains have been trained to adopt classical physical thinking, certainly yes!

It is hardly surprising that some of the great founding fathers of the quantum theory disputed these facts, because they didn't really know the intricacies as to how the quantum world works.

However, it is most certainly true that the consequences of entanglement do and must happen, as has been shown many times as we will see later.

This behaviour is referred to as **non locality quantum behaviour**.

The first experiment to be done to test this, was carried out by the great French physicist Dr. Alain Aspect of the Laboratoire Charles Fabry de l'Institutd'Optique d'Orsay – Institutd'Optique/CNRS/Université Paris Sud 11.

How did he et al carry out this experiment?

Pairs of entangled photons were created using electron decay in calcium atoms.

The photons were immediately sent in opposite directions using the optical fibres in a French telephone system.

The detectors could determine the spin of the quantum particle as soon as it arrived at the detector.

The twin photon always assumed the opposite spin **before light had time to make the journey over the distance**.

Hidden Variable Theories

What are hidden variable theories?

They are theories that imply that our understanding of the quantum world is incomplete, and that the quantum world, per se, is completely deterministic.

In 1927, Count Louis de Broglie, presented a hidden variable theory to the world.

It implied that the quantum world is, in fact, completely deterministic.

The implication is that the only reason that we cannot know a defined quantum state is because there are some hidden variable that, as yet, we don't know about and which are waiting to be discovered.

In other words, our knowledge of the quantum world is deficient and incomplete.

Quantum physicist David Bohm (1917 – 1992) decided to develop de Broglie's hidden variable theories in the 1950s.

De Broglie had reasoned that the trajectory of a quantum particle was determined by a **Pilot Wave**, and was therefore completely deterministic.

This can be represented thus:

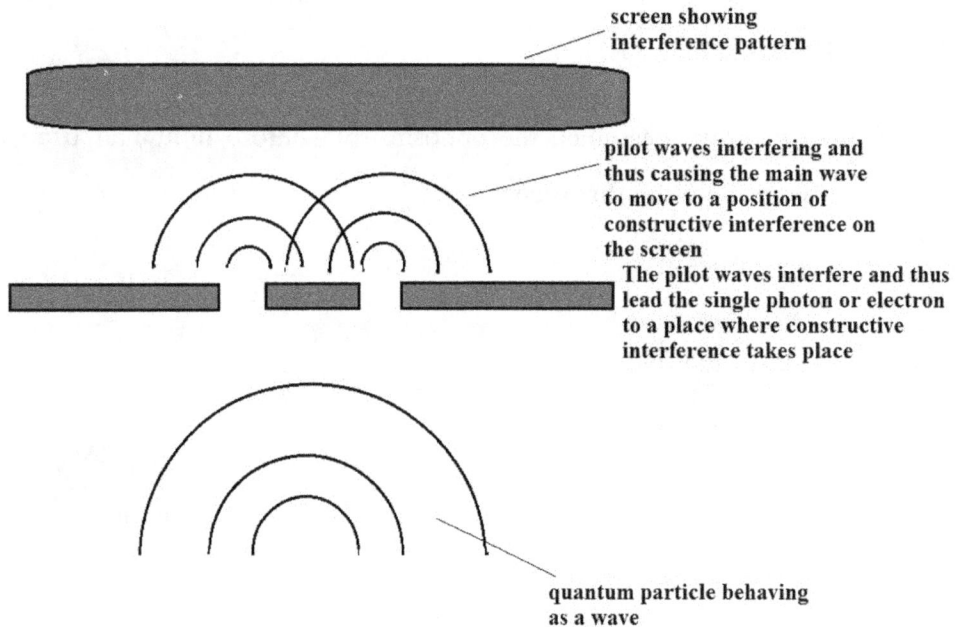

screen showing
interference pattern

pilot waves interfering and thus causing the main wave to move to a position of constructive interference on the screen

The pilot waves interfere and thus lead the single photon or electron to a place where constructive interference takes place

quantum particle behaving as a wave

David Bohm's argument was that the quantum particle's propagation through spacetime was determined by a Ψ **field**, where Ψ is the wave function.

Bohm suggested that the waves never interfered at all!

But that their paths were pre-determined by the Ψ **field**, and that the waves never interfered at all, despite all the evidence to the contrary!!!

Hidden variable theories try to remove the indeterminate nature of the quantum theory, but they are most definitely not widely accepted!

Chapter 28:   The Wave Function and the Universe

Let us now, by way of analogy, consider the workings of an internal combustion engine.

Look at the diagram below, which shows a crankshaft journal and a white metal bearing which is designed to fit **over** the journal.

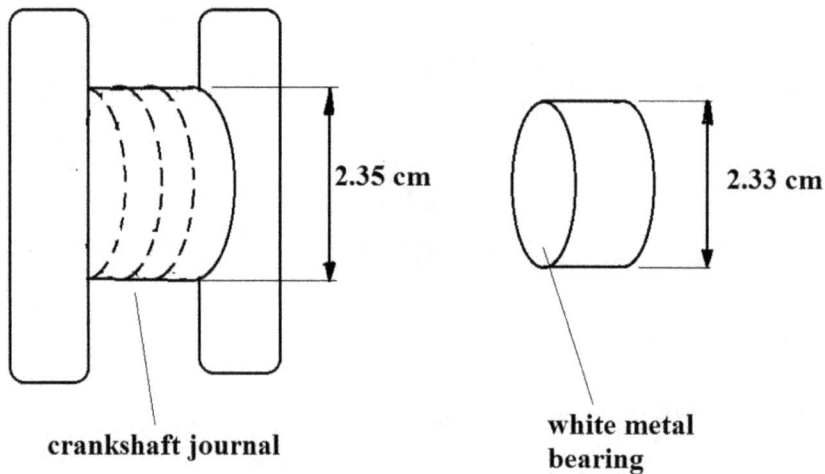

crankshaft journal

white metal bearing

2.35 cm

2.33 cm

Look carefully at the sizes of the bearing, which is designed to fit **over** the cylindrical journal.

What would happen if the bearing shown were to be fitted over the journal shown?

It is quite obvious that the size of the bearing is too small to fit over the journal.

So what would happen if all the journals in the crankshaft had the bearings shown fitted over them and bolted firmly down upon them?

Quite clearly, the engine would not work and the crankshaft would not turn at all because the bearings have been machined to too small a size for the journal shown.

Consider now the piston which fits inside the cylinder on an internal combustion engine.

Look carefully at the diagram below:

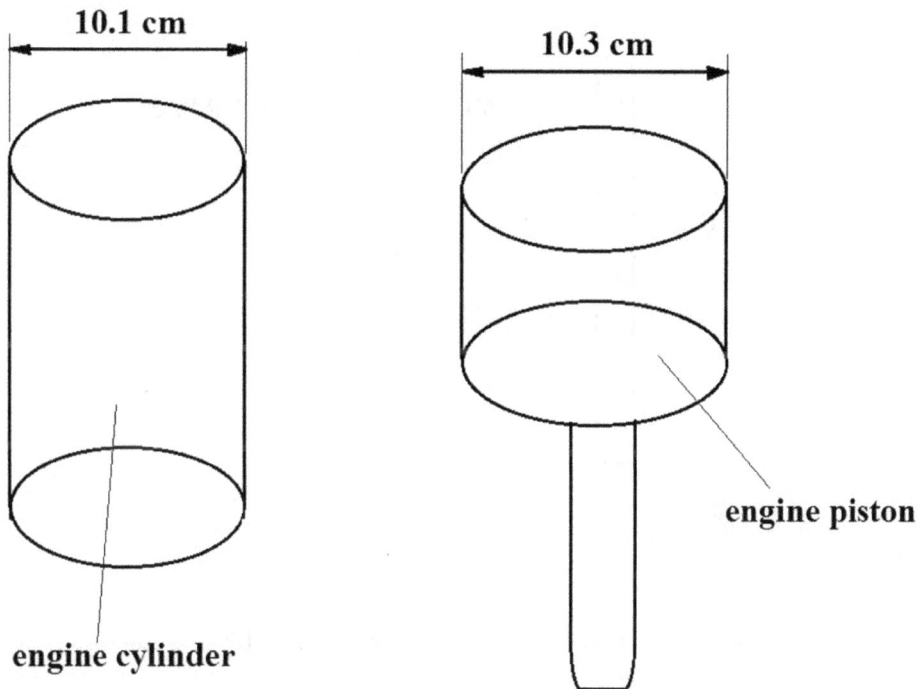

10.1 cm

10.3 cm

engine piston

engine cylinder

What would happen if the piston shown were to be offered into the cylinder shown?

In fact, the piston would not fit at all because it is oversize.

Such a piston would not work in the cylinder shown for this particular internal combustion engine.

What is being shown here:

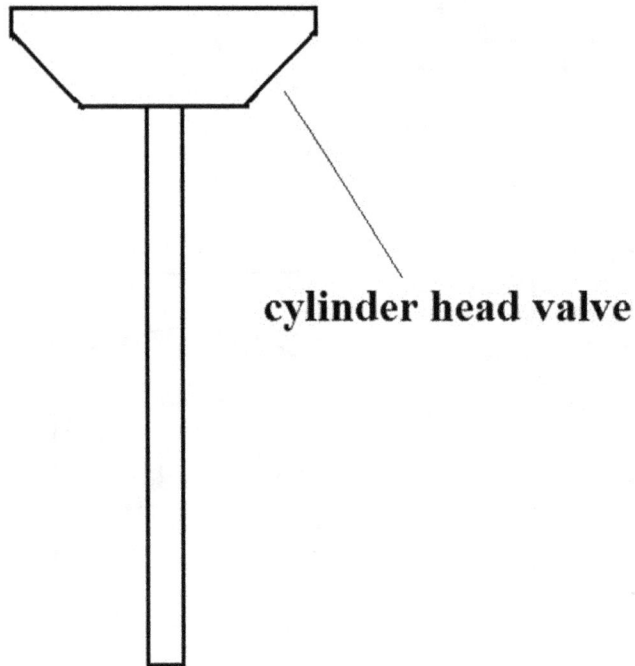

cylinder head valve

This is a cylinder head valve.

What would happen if the valve were to be ground in to be one millimetre larger than the valve seating?

The consequence of this would be that the valve would not sit in the seating with perfect sealing and the high compression fuel air mix would not hold its pressure when the fuel entry cycle had been completed.

As a result, the fuel air mixture would escape and the engine would not fire up.

Our Universe is very similar in principle to the engine.

In which way can our Universe possible similar to the workings of the internal combustion engine?

The answer is that the arrays of fixed physical constants that the Universe runs on have fixed values.

These values in theory could have any value between zero and infinity, but in actual fact, if most of these constants were to differ by quite small amounts from their present values, then a whole array of things that we take for granted would not exist.

For example, if the strength of gravity in the Universe, **G**, were to be just some **6%** less in value, then no galaxies would have formed and we would not be here.

But even more precise than that is the value of the strong nuclear force.

All the carbon 12 that is in our bodies is made in the cores of stars.

Now the average person has some 17 kg of carbon in their bodies, and every bit of it has been made in the cores of stars (usually giant stars).

The only carbon which is made outside of the cores of stars is carbon 14, which is made in the Earth's upper atmosphere, but remember that only one carbon 14 atom exists for every one billion carbon 12 atoms.

If the value of the strong nuclear force were to be reduced by one half of one per cent, then the Hoyle resonance would not exist and no carbon would have ever been formed in star's cores and we, consequently, would not exist at all.

So we see that, in many cases, the values of the physical constants that we have measured around us are very finely tuned, in most cases.

The big question regarding the quantum physics is this:

How did our constants get their finely tuned and serendipitous values that they have?

What do the following represent?

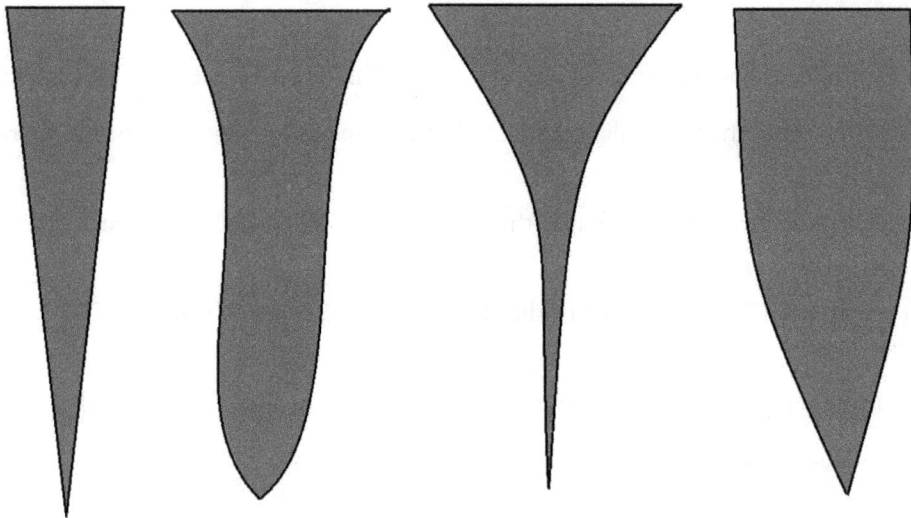

They represent the possible evolution of different universes from the time of the Big Bang to the present.

Some show a constant evolution of spacetime, whilst others show an evolution with different amounts of expansion energy at different times.

However, none of the above fit the now known expansion history of our Universe.

However, we now believe that our Universe evolved according to the following representation, which shows a period of very rapid inflation just after the Big Bang, (courtesy of Professor Alan Guth), followed by a period of more gentle inflation, which we believe to be increasing, (courtesy of Saul Perlmutter and others in 1998).

In 1998, Saul Perlmutter and others decided to look at distant type 1(a) supernovae.

Remember that type 1(a) supernovae are considered to be **standard candles**, that is that they are considered to all explode **with the same brightness**.

A type 1(a) supernova occurs when we have a binary star system in which one of the stars has become a red giant and has, since then shed about forty per cent of its mass into its surroundings and has become a white dwarf.

Remember also that the Chandrasekhar limit is about 1.38 solar masses for the dead white dwarf core.

The process of a type 1(a) supernova is that if the companion star now is also a red giant, then the very high gravity field of the white dwarf can accrete matter from the red giant, and when the mass of the white dwarf finally exceeds the

Chandrasekhar limit, the result is a large explosion, which is believed to be the same brightness each and every time it happens.

Saul Perlmutter and others examined type 1(a) supernovae which were about five billion years old, and the thinking at the time was that the gravity in the Universe would have slowed down the expansion of the galaxies somewhat, such that the brightness of the type 1(a) supernovae examined would have been a little greater than that predicted by the Hubble Law.

However, to the great surprise of the cosmological community, the brightness of the type 1(a) supernovae was **somewhat dimmer** than expected.

What could this possibly mean?

It could only mean that the recession of galaxies had speeded up somewhat, meaning that there had to be a force which was actually accelerating the galaxies apart instead of drawing them together.

This unknown force is now called **dark energy**.

Further, we believe now that the period of gentle inflation which we know happened some billions of years ago, will show an increase in the value of the inflation rate, because we now know that the Universe is actually inflating, but at a more gentle rate than the period of rapid inflation that occurred at about some (1×10^{-34}) of a second after the Big Bang had occurred.

The outline of the history of expansion of our Universe can now be represented thus:

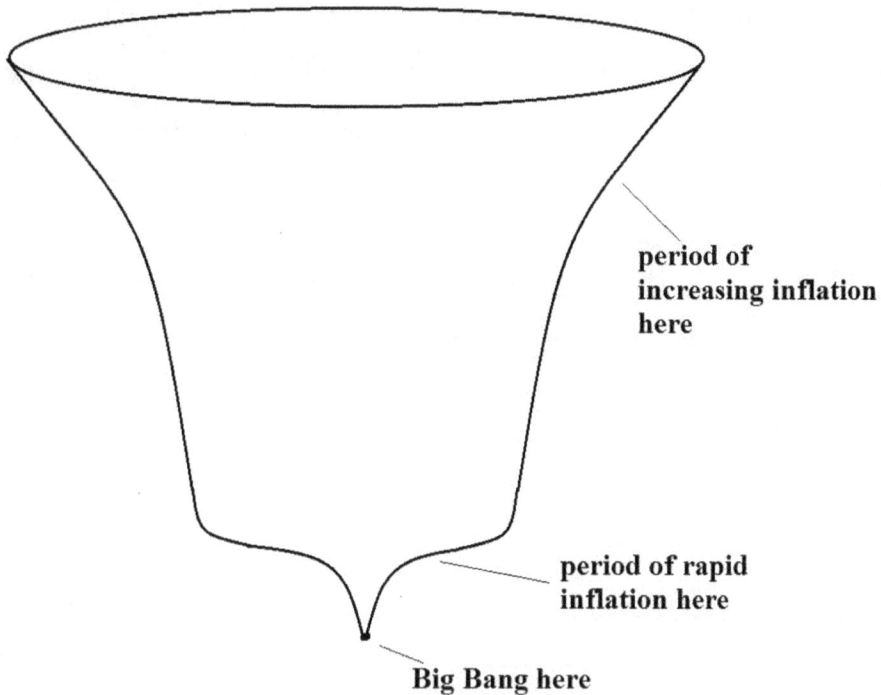

**period of
increasing inflation
here**

**period of rapid
inflation here**

Big Bang here

But what has this got to do with the quantum theory?

Well certain physicists believe that our whole Universe started out as a quantum particle (which might well be true), and that it kept its quantum nature until (so called) intelligent life evolved somewhere, and made an observation on the Universe.

Of course, the "intelligent life" that made the first observation of the Universe might not have been us here on the Earth!

The postulation is that the Universe evolved with **all possible histories**, (in the same vein as Feynman's sum over histories concept).

The observation (which might or might not have emanated from our planet (if there are other intelligent species elsewhere), and consequently the constant values were reduced by way of a wave function collapse, **but only in our part of the Universe**.

Consequently, the Universe continued to evolve according to all the other possible histories, such that we can never ever see those evolutions.

This idea is called **the Flexiverse**.

Do you think that the Universe could be considered to be a quantum system?

Do you think that the Universe could have evolved all possible histories?

What do you think of that proposal?

Chapter 29: Answers to questions

Chapter 3

1.   Energy = 6.6262×10^{-24} Joules

      This is equivalent to 4.1414×10^{-5} eV

      This is insufficient to ionise atomic hydrogen

2.

(a) Energy = 2.4517×19^{-18} Joules

(c) This is equivalent to 15.323eV

(d) This will give a speed of 5.5×10^5 ms^{-1}

3.   (a) Energy = 8.0×10^{-19} Joules

      (b) This is 5.0 eV

      (c) This is 3.2×10^{-19} Joules

      (d) This is 2.0 eV

      (e) The electron will be ionised

      (f) 3.0 eV

      (g) 7.259×10^5 ms^{-1}

Chapter 5

1. $r_2 = 2.1167 \times 10^{-10}$ m, $r_5 = 1.3229 \times 10^{-9}$ m

2. Gap $= 6.35 \times 10^{-10}$ m

3. $r_1 = 5.2918 \times 10^{-11}$ m

Chapter 6

1. Energy $= 5.4497 \times 10^{-19}$ Joules, which is 3.41 eV

2. Energy difference $= 1.05968 \times 10^{-19}$ Joules $= 0.6623$ eV

3. Energy difference $= 1.93769 \times 10^{-18}$ Joules $= 12.11$ eV

Chapter 7

1. Wavelength $= 6.56335 \times 10^{-7}$ which is visible light

2. Wavelength $= 1.2154 \times 10^{-7}$ metre which is ultra violet radiation

Chapter 8

(a) $k = 3.1 \times 10^{10}$ metre^{-1}

(b) Probability $= 1.7 \times 10^{-9}$

Chapter 13

1. de Broglie wavelength $= 2.2087 \times 10^{-34}$ metre

2. de Broglie wavelength $= 2.2087 \times 10^{-36}$ metre – this is impossible since it is less than the Planck length

Chapter 15

1.   (a) 2.5×10^{-6} m

(c) 4.0×10^{-7} m

(d)   Colour is blue

2.

(a) 2128 lines per cm.

(b) Colour is approximately blue

Chapter 21

1. $\Delta x = 0.03148$ metre

2. $\Delta x = 28.9419$ metre

Chapter 25

1. Energy available $= 978.1$ MJ

2. Energy to produce a neutron / antineutron pair for less than the Planck time is

calculated from the Einstein equation $\mathbf{E = m c^2}$.

Energy $= 3.0107 \times 10^{-10}$ Joules

This means that the energy has an excess of some 3.249×10^{18} times!

Chapter 26

1. 6.3131×10^{-31} metre

<u>References</u>

(1) "To infinity and beyond", by E. Maor (Princeton 1991)

(2) Quoted from: "The Many Worlds of Hugh Everett 3rd" by Peter Byrne p.325

Index

www.ingramcontent.com/pod-product-compliance
Lightning Source LLC
Chambersburg PA
CBHW080553220326
41599CB00032B/6468